Assessments and Classroom Masters

SAXON MATH™ 7/6

Stephen Hake
John Saxon

SAXON™
PUBLISHERS

Saxon Publishers gratefully acknowledges the contributions of the following individuals in the completion of this project:

Authors: Stephen Hake, John Saxon

Editorial: Brian E. Rice, Matt Maloney, R. Clint Keele, Andrew C. Kershen, Sean Douglas

Editorial Support Services: Christopher Davey, Susan Toth, Darlene C. Terry, Jack Day, Cassandra Phillips

Production: Sunnie Atkins, Lois Rossman, Julie A. Webb, Mary Sue Butner, Cristi D. Whiddon

Project Management: Angela Johnson, Becky Cavnar

Saxon Publishers would also like to thank Dan Shippey of Delta 7 Studios for the model reproduced on Activity Masters 12 and 13.

Printed in the United States of America

ISBN: 1-59141-288-9

Manufacturing Code: 1 2 3 4 5 6 7 8 022 12 11 10 09 08 07 06 05

CONTENTS

Masters

Answers and Solutions

Introduction

Saxon Math 7/6 Assessments and Classroom Masters contains Facts Practice Tests, Course Assessments, Tests, Test-Day Activity Masters, Activity Masters, and Forms, as well as answers and solutions. Brief descriptions of these components are provided below. Additional information can be found on the pages that introduce each section.

About the Facts Practice Tests

Facts Practice Tests are an essential and integral part of the daily program. Mastery of basic facts frees students to focus on procedures and concepts rather than computation. Employing memory to recall frequently encountered facts permits students to bring higher-level thinking skills to bear when solving problems.

Facts Practice Tests should be administered before every lesson and test as prescribed in the Teacher's Manual. One copy of the suggested facts practice should be given to each student.† Limit student work on these tests to five minutes or less.

Facts Practice Tests O–Q are not mentioned in conjunction with any lesson or test. They may be substituted for prescribed Facts Practice Tests whose content students have obviously mastered.

About the Course Assessments

A multiple-choice Pretest and Posttest for *Saxon Math 7/6* is included. Administer this test once early in the school year to gauge the skills of incoming students and again late in the school year to determine growth in student competency. Some of the content of this test might be unfamiliar to students initially; however, most of the content covers skills and concepts that are included in the math curriculum of the preceding year. Test results should show growing mastery of the foundational knowledge necessary for continued success.

The Final Exam is a comprehensive multiple-choice test that assesses student knowledge of the content presented during the course; it should be administered as late in the year as possible.

About the Tests

The tests are designed to reward students and to provide teachers with diagnostic information. Every lesson in the student textbook culminates with a cumulative mixed practice, so the tests are cumulative as well. The test design allows students to display the skills they have developed, and it fosters confidence that will benefit students when they encounter comprehensive standardized tests. Cumulative assessment also gives students an incentive to master skills and concepts that might be troublesome initially.

The testing schedule is built into the Teacher's Manual. In addition, an overview of the schedule is printed in this book on the page immediately preceding the first test. Administering the tests according to the schedule is essential. Each test is written to follow a specific five-lesson interval in the textbook. Following the schedule allows students sufficient practice on new topics before they are assessed on those topics.

Two forms of each test are provided. We recommend that the second form be used for makeup tests if needed.

---

†As an alternative to copying a Facts Practice Test for each student every day, teachers may provide each student a *Saxon Math 7/6 Facts Practice Workbook,* which contains all the facts practices recommended for one student for one year.

About the Test-Day Activity Masters

A Test-Day Activity is a brief exercise that takes advantage of the math class time that remains after students complete their tests. The Teacher's Manual recommends one for use after each test. Most of these activities are supported by worksheets called Test-Day Activity Masters, which should be copied from this book as directed in the Teacher's Manual.

About the Activity Masters

Selected lessons and investigations in the student textbook present content through activities. Often these activities require the use of worksheets called Activity Masters that must be copied for each student or group of students. The Teacher's Manual indicates when to photocopy these masters and how many copies to make.

About the Forms

This book contains two kinds of forms teachers might find useful. The Answer Forms provide sufficient space for students to record their work on daily assignments and tests. The Recording Forms are designed to help teachers track and analyze student performance on tests.

Parent Letter

Parental involvement greatly improves a student's chance for success. When parents understand what is being asked of their children in school, they can provide support at home. The following parent letter explains the keys to success in Saxon mathematics programs. An English and a Spanish version are provided. At the beginning of the year, give each student one copy to take to his or her parents. To ensure that the parents receive the letters, you might require the students to return the letters bearing their parents' signatures. Alternatively, you could mail the letters directly to the parents.

Dear Parent or Guardian:

This year your child will be using a mathematics textbook from Saxon Publishers. Saxon books are unlike the traditional texts most of us used in the past. You may remember that in those traditional texts students were taught and expected to learn an entire mathematical concept in one day. The homework for that day consisted of twenty or thirty problems, all dealing with the same concept.

In contrast, Saxon textbooks divide concepts into smaller, more easily grasped pieces called increments. A new increment is presented each day, and students work only a few problems each day involving the new material. The remaining homework problems cover previously introduced concepts. Thus, every assignment (and every test) is a cumulative review of material covered up to that point.

This is where your involvement is crucial. It is essential that your child complete each day's assignment. You can expect your child to complete three to five lessons per week with a test after every fifth lesson. Please ask to see your child's daily assignment, and encourage him or her to complete all the work. Students who complete all the daily assignments will have much to be proud of at test time, so you will have much to praise.

Some key words in the preface of your child's textbook are printed in boldface. They remind your child of the importance of the daily exercises. Those words are worth repeating here.

> **Solving these problems day after day is the secret to success. Work each problem in every practice set, in every problem set, and in every investigation. Do not skip problems. With honest effort you will experience success and true learning, which will stay with you and serve you well in the future.**

Mathematics is not difficult; mathematics is just different. Tasks that are different become familiar through practice. Learning mathematics, like learning a foreign language, a musical instrument, or an athletic skill, requires long-term practice to develop and maintain mastery of the requisite skills.

The study of mathematics can be either exciting and rewarding or dull and frustrating. Your child's experience in mathematics will be a direct result of his or her attitude and effort. A positive attitude and a commitment to long-term practice greatly improve the chances of your child progressing into higher-level mathematics courses. Through our combined efforts your child can experience success at this level and learn the foundational skills needed for the next level.

Sincerely,

Estimados padres o tutores:

Este año su hijo(a) usará un texto de matemáticas de Saxon Publishers. Los libros de Saxon no son como los textos tradicionales que acostumbrábamos a usar en el pasado. Quizás ustedes recuerden que en esos textos tradicionales se esperaba que los estudiantes conocieran y aprendieran un concepto en un solo día. La tarea para ese día consistía en veinte o treinta problemas que trataban sobre el mismo concepto.

Por el contrario, los libros de texto de Saxon dividen los conceptos en unidades más pequeñas y más fáciles de comprender. Un nuevo incremento es introducido cada día y cada día los estudiantes sólo resuelven unos pocos problemas que tratan sobre este nuevo material. El resto de la tarea consiste en la práctica de problemas basados en conceptos presentados previamente. Cada asignación (como así mismo cada examen) ofrece un repaso acumulativo del material cubierto hasta ese momento.

Es en este punto donde la participación de ustedes es crítica. Pueden contar con que su hijo(a) completará 3–5 lecciones por semana y tomará un examen después de cada cinco lecciones. Por favor, pídale a su hijo(a) que le muestre la tarea todos los días y anímele a que la haga en su totalidad. Los estudiantes que hayan completado todas las asignaciones diarias se sentirán muy orgullosos del resultado de sus exámenes y ustedes tendrán muchas razones para felicitarlos.

Unas palabras claves del prólogo del libro de su hijo(a) están impresas en negrita. Sirven para recordar a su hijo(a) la importancia de los ejercicios diarios. Vale la pena repetir este texto aquí:

> **Solucionar estos problemas día a día es el secreto del éxito. Resuelve cada problema en las secciones de Práctica y de Problemas y en cada Investigación. No te saltes ninguno. A través del esfuerzo honesto alcanzarás el éxito y obtendrás conocimientos duraderos que se te quedarán y que te serán muy útiles en el futuro.**

Las matemáticas no son difíciles; son simplemente diferentes. Todo aquello que es diferente se vuelve familiar a través de la práctica. El aprendizaje de las matemáticas, al igual que el aprendizaje de un nuevo idioma, el de un instrumento musical o el de un deporte, requiere extensa práctica para desarrollar y mantener el dominio de las destrezas requeridas.

El estudio de las matemáticas puede ser o emocionante y gratificante o aburrido y frustrante. La experiencia de su hijo(a) será el resultado directo de su actitud y su esfuerzo. Una actitud positiva y un compromiso a la práctica a largo plazo aumentan mucho las posibilidades de que su hijo(a) progrese hasta los cursos de matemáticas avanzadas. Gracias al esfuerzo de todos nosotros, su hijo(a) puede lograr el éxito en este nivel y aprender los conocimientos fundamentales que necesitará para tener éxito en el futuro.

Sinceramente,

Facts Practice Tests

Rapid and accurate recall of basic facts and skills dramatically increases students' mathematical powers. To that end we have provided the following series of Facts Practice Tests. Reproduce these pages for each student.†

Begin each lesson with the Facts Practice Test suggested in the Warm-Up, and limit the practice to five minutes or less. On the first presentation of a new practice sheet, you may choose to make the practice instructional by working through the problems with the students and providing the answers. However, students should work independently and rapidly during all subsequent practices of the same test, racing to improve on previous performances.

Each Facts Practice Test contains a line for students to record their times. Timing students is motivating. Striving to improve speed helps to automate skills and offers the additional benefit of an up-tempo atmosphere for the class. Time invested in facts practice is repaid in the students' ability to work faster.

Quickly read answers at the conclusion of the practice. (The answers are found in the back of this book.) Students should correct errors and complete the sheet as part of the day's assignment if they are unable to finish within the allotted time.

On test day students should be held accountable for mastering the content of recent facts practice sessions. Each test identifies a Facts Practice Test to be taken on that day. Allow five minutes on test days for students to complete their Facts Practice Test.

Additional, optional Facts Practice Tests that are not specifically suggested in the Warm-Ups are included in this section. These tests may be used as alternate or supplemental facts practice activities at the teacher's discretion.

---

†As an alternative to copying a Facts Practice Test for each student every day, provide each student a *Saxon Math 7/6 Facts Practice Workbook,* which contains all the facts practices recommended for one student for one year.

F A C T S   P R A C T I C E   T E S T

A

64 Addition Facts

Name ______________________

Time ______________________

Add.

$\begin{array}{r} 7 \\ +\ 2 \\ \hline \end{array}$	$\begin{array}{r} 9 \\ +\ 4 \\ \hline \end{array}$	$\begin{array}{r} 2 \\ +\ 8 \\ \hline \end{array}$	$\begin{array}{r} 6 \\ +\ 5 \\ \end{array}$	$\begin{array}{r} 4 \\ +\ 4 \\ \hline \end{array}$	$\begin{array}{r} 3 \\ +\ 9 \\ \hline \end{array}$	$\begin{array}{r} 8 \\ +\ 4 \\ \hline \end{array}$	$\begin{array}{r} 5 \\ +\ 7 \\ \hline \end{array}$
$\begin{array}{r} 9 \\ +\ 7 \\ \hline \end{array}$	$\begin{array}{r} 4 \\ +\ 7 \\ \hline \end{array}$	$\begin{array}{r} 7 \\ +\ 5 \\ \hline \end{array}$	$\begin{array}{r} 5 \\ +\ 4 \\ \hline \end{array}$	$\begin{array}{r} 3 \\ +\ 4 \\ \hline \end{array}$	$\begin{array}{r} 6 \\ +\ 8 \\ \hline \end{array}$	$\begin{array}{r} 2 \\ +\ 5 \\ \hline \end{array}$	$\begin{array}{r} 8 \\ +\ 8 \\ \hline \end{array}$
$\begin{array}{r} 6 \\ +\ 3 \\ \hline \end{array}$	$\begin{array}{r} 2 \\ +\ 9 \\ \hline \end{array}$	$\begin{array}{r} 7 \\ +\ 8 \\ \hline \end{array}$	$\begin{array}{r} 8 \\ +\ 3 \\ \hline \end{array}$	$\begin{array}{r} 5 \\ +\ 9 \\ \hline \end{array}$	$\begin{array}{r} 3 \\ +\ 6 \\ \hline \end{array}$	$\begin{array}{r} 9 \\ +\ 9 \\ \hline \end{array}$	$\begin{array}{r} 4 \\ +\ 9 \\ \hline \end{array}$
$\begin{array}{r} 5 \\ +\ 8 \\ \hline \end{array}$	$\begin{array}{r} 9 \\ +\ 5 \\ \hline \end{array}$	$\begin{array}{r} 4 \\ +\ 5 \\ \hline \end{array}$	$\begin{array}{r} 8 \\ +\ 6 \\ \hline \end{array}$	$\begin{array}{r} 2 \\ +\ 3 \\ \hline \end{array}$	$\begin{array}{r} 6 \\ +\ 6 \\ \hline \end{array}$	$\begin{array}{r} 5 \\ +\ 2 \\ \hline \end{array}$	$\begin{array}{r} 7 \\ +\ 3 \\ \hline \end{array}$
$\begin{array}{r} 3 \\ +\ 8 \\ \hline \end{array}$	$\begin{array}{r} 8 \\ +\ 9 \\ \hline \end{array}$	$\begin{array}{r} 2 \\ +\ 2 \\ \hline \end{array}$	$\begin{array}{r} 7 \\ +\ 6 \\ \hline \end{array}$	$\begin{array}{r} 5 \\ +\ 5 \\ \hline \end{array}$	$\begin{array}{r} 6 \\ +\ 9 \\ \hline \end{array}$	$\begin{array}{r} 3 \\ +\ 7 \\ \hline \end{array}$	$\begin{array}{r} 9 \\ +\ 8 \\ \hline \end{array}$
$\begin{array}{r} 4 \\ +\ 2 \\ \hline \end{array}$	$\begin{array}{r} 3 \\ +\ 3 \\ \hline \end{array}$	$\begin{array}{r} 6 \\ +\ 4 \\ \hline \end{array}$	$\begin{array}{r} 4 \\ +\ 8 \\ \hline \end{array}$	$\begin{array}{r} 9 \\ +\ 3 \\ \hline \end{array}$	$\begin{array}{r} 2 \\ +\ 4 \\ \hline \end{array}$	$\begin{array}{r} 8 \\ +\ 5 \\ \hline \end{array}$	$\begin{array}{r} 7 \\ +\ 9 \\ \hline \end{array}$
$\begin{array}{r} 7 \\ +\ 4 \\ \hline \end{array}$	$\begin{array}{r} 2 \\ +\ 6 \\ \hline \end{array}$	$\begin{array}{r} 5 \\ +\ 3 \\ \hline \end{array}$	$\begin{array}{r} 9 \\ +\ 6 \\ \hline \end{array}$	$\begin{array}{r} 4 \\ +\ 3 \\ \hline \end{array}$	$\begin{array}{r} 6 \\ +\ 7 \\ \hline \end{array}$	$\begin{array}{r} 3 \\ +\ 2 \\ \hline \end{array}$	$\begin{array}{r} 8 \\ +\ 7 \\ \hline \end{array}$
$\begin{array}{r} 5 \\ +\ 6 \\ \hline \end{array}$	$\begin{array}{r} 8 \\ +\ 2 \\ \hline \end{array}$	$\begin{array}{r} 3 \\ +\ 5 \\ \hline \end{array}$	$\begin{array}{r} 6 \\ +\ 2 \\ \hline \end{array}$	$\begin{array}{r} 7 \\ +\ 7 \\ \hline \end{array}$	$\begin{array}{r} 4 \\ +\ 6 \\ \hline \end{array}$	$\begin{array}{r} 9 \\ +\ 2 \\ \hline \end{array}$	$\begin{array}{r} 2 \\ +\ 7 \\ \hline \end{array}$

B

100 Addition Facts

Name ______________________

Time ______________________

Add.

3 + 2	8 + 3	2 + 1	5 + 6	2 + 9	4 + 8	8 + 0	3 + 9	1 + 0	6 + 3
7 + 3	1 + 6	4 + 7	0 + 3	6 + 4	5 + 5	3 + 1	7 + 2	8 + 5	2 + 5
4 + 0	5 + 7	1 + 1	5 + 4	2 + 8	7 + 1	4 + 6	0 + 2	6 + 5	4 + 9
8 + 6	0 + 4	5 + 8	7 + 4	1 + 7	6 + 6	4 + 1	8 + 2	2 + 4	6 + 0
9 + 1	8 + 8	2 + 2	4 + 5	6 + 2	0 + 0	5 + 9	3 + 3	8 + 1	2 + 7
4 + 4	7 + 5	0 + 1	8 + 7	3 + 4	7 + 9	1 + 2	6 + 7	0 + 8	9 + 2
0 + 9	8 + 9	7 + 6	1 + 3	6 + 8	2 + 0	8 + 4	3 + 5	9 + 8	5 + 0
9 + 3	2 + 6	3 + 0	6 + 1	3 + 6	5 + 2	0 + 5	6 + 9	1 + 8	9 + 6
4 + 3	9 + 9	0 + 7	9 + 4	7 + 7	1 + 4	3 + 7	7 + 0	2 + 3	5 + 1
9 + 5	1 + 5	9 + 0	3 + 8	1 + 9	5 + 3	4 + 2	9 + 7	0 + 6	7 + 8

C

100 Subtraction Facts

Name ______________________

Time ______________________

Subtract.

16 − 9	7 − 1	18 − 9	11 − 3	13 − 7	8 − 2	11 − 5	5 − 0	17 − 9	6 − 1
10 − 9	6 − 2	13 − 9	4 − 0	10 − 5	5 − 1	10 − 3	12 − 6	10 − 1	6 − 4
7 − 2	14 − 7	8 − 1	11 − 6	3 − 3	16 − 7	5 − 2	12 − 4	3 − 0	11 − 7
17 − 8	6 − 0	10 − 6	4 − 1	9 − 5	9 − 0	5 − 4	12 − 5	4 − 2	9 − 3
12 − 3	16 − 8	9 − 1	15 − 6	11 − 4	13 − 5	1 − 0	8 − 5	9 − 6	11 − 2
7 − 0	10 − 8	6 − 3	14 − 5	3 − 1	8 − 6	4 − 4	11 − 8	3 − 2	15 − 9
13 − 8	7 − 4	10 − 7	0 − 0	12 − 8	5 − 5	4 − 3	8 − 7	7 − 3	7 − 6
5 − 3	7 − 5	2 − 1	6 − 6	8 − 4	2 − 2	13 − 6	15 − 8	2 − 0	13 − 9
1 − 1	11 − 9	10 − 4	9 − 2	14 − 6	8 − 0	9 − 4	10 − 2	6 − 5	8 − 3
7 − 7	14 − 8	12 − 9	9 − 8	12 − 7	9 − 9	15 − 7	8 − 8	14 − 9	9 − 7

FACTS PRACTICE TEST

D

64 Multiplication Facts

Name ______________________

Time ______________________

Multiply.

5×6	4×3	9×8	7×5	2×9	8×4	9×3	6×9
9×4	2×5	7×6	4×8	7×9	5×4	3×2	9×7
3×7	8×5	6×2	5×5	3×5	2×4	7×7	8×9
6×4	2×8	4×4	8×2	3×9	6×6	9×9	5×3
4×6	8×8	5×7	6×3	2×2	7×4	3×8	8×6
2×6	5×9	3×3	9×2	6×7	4×5	7×2	9×6
5×2	7×8	2×3	6×8	4×7	9×5	3×6	8×7
3×4	7×3	5×8	4×2	8×3	2×7	6×5	4×9

FACTS PRACTICE TEST

E

100 Multiplication Facts

Name ______________________

Time ______________________

Multiply.

9×9	3×5	8×5	2×6	4×7	0×3	7×2	1×5	7×8	4×0
3×4	5×9	0×2	7×3	4×1	2×7	6×3	5×4	1×0	9×2
1×1	9×0	2×8	6×4	0×7	8×1	3×3	4×8	9×3	2×0
4×9	7×0	1×2	8×4	6×5	2×9	9×4	0×1	7×4	5×8
0×8	4×2	9×8	3×6	5×5	1×6	5×0	6×6	2×1	7×9
9×1	2×2	5×1	4×3	0×0	8×9	3×7	9×7	1×7	6×0
5×6	7×5	3×0	8×8	1×3	8×3	5×2	0×4	9×5	6×7
2×3	8×6	0×5	6×1	3×8	7×6	1×8	9×6	4×4	5×3
7×7	1×4	6×2	4×5	2×4	8×0	3×1	6×8	0×9	8×7
3×2	4×6	1×9	5×7	8×2	0×6	7×1	2×5	6×9	3×9

FACTS PRACTICE TEST

F

90 Division Facts

Name ______________________

Time ______________________

Divide.

$7\overline{)21}$	$2\overline{)10}$	$6\overline{)42}$	$1\overline{)3}$	$4\overline{)24}$	$3\overline{)6}$	$9\overline{)54}$	$6\overline{)18}$	$4\overline{)0}$	$5\overline{)30}$
$4\overline{)32}$	$8\overline{)56}$	$1\overline{)0}$	$6\overline{)12}$	$3\overline{)18}$	$9\overline{)72}$	$5\overline{)15}$	$2\overline{)8}$	$7\overline{)42}$	$6\overline{)36}$
$6\overline{)0}$	$5\overline{)10}$	$9\overline{)9}$	$2\overline{)6}$	$7\overline{)63}$	$4\overline{)16}$	$8\overline{)48}$	$1\overline{)2}$	$5\overline{)35}$	$3\overline{)21}$
$2\overline{)18}$	$6\overline{)6}$	$3\overline{)15}$	$8\overline{)40}$	$2\overline{)0}$	$5\overline{)20}$	$9\overline{)27}$	$1\overline{)8}$	$4\overline{)4}$	$7\overline{)35}$
$4\overline{)20}$	$9\overline{)63}$	$1\overline{)4}$	$7\overline{)14}$	$3\overline{)3}$	$8\overline{)24}$	$5\overline{)0}$	$6\overline{)24}$	$8\overline{)8}$	$2\overline{)16}$
$5\overline{)5}$	$8\overline{)64}$	$3\overline{)0}$	$4\overline{)28}$	$7\overline{)49}$	$2\overline{)4}$	$9\overline{)81}$	$3\overline{)12}$	$6\overline{)30}$	$1\overline{)5}$
$8\overline{)32}$	$1\overline{)1}$	$9\overline{)36}$	$3\overline{)27}$	$2\overline{)14}$	$5\overline{)25}$	$6\overline{)48}$	$8\overline{)0}$	$7\overline{)28}$	$4\overline{)36}$
$2\overline{)12}$	$5\overline{)45}$	$1\overline{)7}$	$4\overline{)8}$	$7\overline{)0}$	$8\overline{)16}$	$3\overline{)24}$	$9\overline{)45}$	$1\overline{)9}$	$6\overline{)54}$
$7\overline{)56}$	$9\overline{)0}$	$8\overline{)72}$	$2\overline{)2}$	$5\overline{)40}$	$3\overline{)9}$	$9\overline{)18}$	$1\overline{)6}$	$4\overline{)12}$	$7\overline{)7}$

FACTS PRACTICE TEST

G

30 Fractions to Reduce

Cut copies into two tests.

Test G—30 Fractions to Reduce

Name ________________ Time ________

Reduce each fraction to lowest terms.

$\frac{2}{8}$ =	$\frac{4}{6}$ =	$\frac{6}{10}$ =	$\frac{2}{4}$ =	$\frac{6}{16}$ =
$\frac{5}{100}$ =	$\frac{9}{12}$ =	$\frac{14}{16}$ =	$\frac{4}{10}$ =	$\frac{4}{12}$ =
$\frac{2}{10}$ =	$\frac{3}{6}$ =	$\frac{25}{100}$ =	$\frac{3}{12}$ =	$\frac{4}{16}$ =
$\frac{3}{9}$ =	$\frac{10}{16}$ =	$\frac{6}{9}$ =	$\frac{4}{8}$ =	$\frac{2}{12}$ =
$\frac{6}{12}$ =	$\frac{2}{16}$ =	$\frac{8}{10}$ =	$\frac{2}{6}$ =	$\frac{75}{100}$ =
$\frac{12}{16}$ =	$\frac{8}{12}$ =	$\frac{6}{8}$ =	$\frac{10}{12}$ =	$\frac{5}{10}$ =

Test G—30 Fractions to Reduce

Name ________________ Time ________

Reduce each fraction to lowest terms.

$\frac{2}{8}$ =	$\frac{4}{6}$ =	$\frac{6}{10}$ =	$\frac{2}{4}$ =	$\frac{6}{16}$ =
$\frac{5}{100}$ =	$\frac{9}{12}$ =	$\frac{14}{16}$ =	$\frac{4}{10}$ =	$\frac{4}{12}$ =
$\frac{2}{10}$ =	$\frac{3}{6}$ =	$\frac{25}{100}$ =	$\frac{3}{12}$ =	$\frac{4}{16}$ =
$\frac{3}{9}$ =	$\frac{10}{16}$ =	$\frac{6}{9}$ =	$\frac{4}{8}$ =	$\frac{2}{12}$ =
$\frac{6}{12}$ =	$\frac{2}{16}$ =	$\frac{8}{10}$ =	$\frac{2}{6}$ =	$\frac{75}{100}$ =
$\frac{12}{16}$ =	$\frac{8}{12}$ =	$\frac{6}{8}$ =	$\frac{10}{12}$ =	$\frac{5}{10}$ =

FACTS PRACTICE TEST

H

72 Multiplication and Division Facts

Name ______________________

Time ______________________

Multiply or divide as indicated.

5×9	$6\overline{)36}$	4×7	$8\overline{)40}$	10×3	$5\overline{)20}$	2×7	$9\overline{)27}$
$9\overline{)81}$	5×7	$6\overline{)24}$	2×3	$7\overline{)42}$	7×9	$5\overline{)10}$	4×5
2×9	$7\overline{)28}$	6×9	$4\overline{)8}$	3×3	$8\overline{)24}$	10×4	$6\overline{)12}$
$9\overline{)63}$	2×5	$8\overline{)64}$	7×6	$10\overline{)100}$	3×5	$9\overline{)36}$	10×8
4×8	$5\overline{)25}$	3×8	$4\overline{)16}$	10×5	$7\overline{)14}$	2×2	$8\overline{)48}$
$9\overline{)54}$	3×6	$5\overline{)15}$	2×8	$7\overline{)49}$	4×6	$6\overline{)30}$	5×8
10×6	$4\overline{)12}$	7×8	$6\overline{)18}$	3×4	$7\overline{)35}$	5×6	$8\overline{)32}$
$8\overline{)56}$	2×4	$8\overline{)16}$	6×8	$7\overline{)21}$	8×9	$3\overline{)6}$	3×9
10×7	$9\overline{)18}$	3×7	$9\overline{)45}$	2×6	$9\overline{)72}$	4×9	$9\overline{)90}$

I

28 Improper Fractions to Simplify

Cut copies into two tests.

Test I—28 Improper Fractions to Simplify

Name ______________ Time ______________

Write each improper fraction as a mixed number or a whole number.

$\frac{5}{4}$ =	$\frac{16}{12}$ =	$\frac{12}{8}$ =	$\frac{8}{6}$ =
$\frac{12}{6}$ =	$\frac{12}{10}$ =	$\frac{6}{4}$ =	$\frac{20}{12}$ =
$\frac{5}{3}$ =	$\frac{10}{8}$ =	$\frac{25}{10}$ =	$\frac{10}{3}$ =
$\frac{15}{10}$ =	$\frac{3}{2}$ =	$\frac{9}{6}$ =	$\frac{7}{4}$ =
$\frac{18}{12}$ =	$\frac{8}{3}$ =	$\frac{15}{6}$ =	$\frac{14}{4}$ =
$\frac{8}{4}$ =	$\frac{10}{6}$ =	$\frac{5}{2}$ =	$\frac{21}{12}$ =
$\frac{15}{12}$ =	$\frac{10}{4}$ =	$\frac{15}{8}$ =	$\frac{4}{3}$ =

Test I—28 Improper Fractions to Simplify

Name ______________ Time ______________

Write each improper fraction as a mixed number or a whole number.

$\frac{5}{4}$ =	$\frac{16}{12}$ =	$\frac{12}{8}$ =	$\frac{8}{6}$ =
$\frac{12}{6}$ =	$\frac{12}{10}$ =	$\frac{6}{4}$ =	$\frac{20}{12}$ =
$\frac{5}{3}$ =	$\frac{10}{8}$ =	$\frac{25}{10}$ =	$\frac{10}{3}$ =
$\frac{15}{10}$ =	$\frac{3}{2}$ =	$\frac{9}{6}$ =	$\frac{7}{4}$ =
$\frac{18}{12}$ =	$\frac{8}{3}$ =	$\frac{15}{6}$ =	$\frac{14}{4}$ =
$\frac{8}{4}$ =	$\frac{10}{6}$ =	$\frac{5}{2}$ =	$\frac{21}{12}$ =
$\frac{15}{12}$ =	$\frac{10}{4}$ =	$\frac{15}{8}$ =	$\frac{4}{3}$ =

FACTS PRACTICE TEST

J

24 Mixed Numbers to Write as Improper Fractions

Cut copies into two tests.

Test J—24 Mixed Numbers to Write as Improper Fractions

Name ______________________ Time ______

Write each mixed number as an improper fraction.

$2\frac{1}{2} =$	$2\frac{2}{5} =$	$1\frac{3}{4} =$	$2\frac{3}{4} =$
$2\frac{1}{8} =$	$1\frac{2}{3} =$	$10\frac{1}{2} =$	$2\frac{1}{3} =$
$3\frac{1}{2} =$	$1\frac{5}{6} =$	$2\frac{1}{4} =$	$1\frac{1}{8} =$
$5\frac{1}{2} =$	$1\frac{3}{8} =$	$5\frac{1}{3} =$	$3\frac{1}{4} =$
$4\frac{1}{2} =$	$1\frac{7}{8} =$	$2\frac{2}{3} =$	$3\frac{3}{10} =$
$1\frac{5}{8} =$	$3\frac{3}{4} =$	$2\frac{3}{8} =$	$7\frac{1}{2} =$

Test J— 24 Mixed Numbers to Write as Improper Fractions

Name ______________________ Time ______

Write each mixed number as an improper fraction.

$2\frac{1}{2} =$	$2\frac{2}{5} =$	$1\frac{3}{4} =$	$2\frac{3}{4} =$
$2\frac{1}{8} =$	$1\frac{2}{3} =$	$10\frac{1}{2} =$	$2\frac{1}{3} =$
$3\frac{1}{2} =$	$1\frac{5}{6} =$	$2\frac{1}{4} =$	$1\frac{1}{8} =$
$5\frac{1}{2} =$	$1\frac{3}{8} =$	$5\frac{1}{3} =$	$3\frac{1}{4} =$
$4\frac{1}{2} =$	$1\frac{7}{8} =$	$2\frac{2}{3} =$	$3\frac{3}{10} =$
$1\frac{5}{8} =$	$3\frac{3}{4} =$	$2\frac{3}{8} =$	$7\frac{1}{2} =$

FACTS PRACTICE TEST

K Linear Measurement

Name ____________

Time ____________

Write the abbreviation for each of the following units.

Metric Units

1. millimeter ______

2. centimeter ______

3. meter ______

4. kilometer ______

U.S. Customary Units

5. inch ______

6. foot ______

7. yard ______

8. mile ______

Complete each unit conversion.

Metric Conversions

9. 1 centimeter = ______ millimeters

10. 1 meter = ______ millimeters

11. 1 meter = ______ centimeters

12. 1 kilometer = ______ meters

U.S. Customary Conversions

13. 1 foot = ______ inches

14. 1 yard = ______ inches

15. 1 yard = ______ feet

16. 1 mile = ______ feet

17. 1 mile = ______ yards

Conversions between systems

18. 1 inch = ______ centimeters

19. 1 mile ≈ ______ meters

20. 1 meter ≈ ______ inches

21. 1 kilometer ≈ ______ mile

Write an appropriate unit for each physical reference.

Metric Units

22. The thickness of a dime: ______

23. The width of a little finger: ______

24. The length of one BIG step: ______

U.S. Customary Units

25. The width of two fingers: ______

26. The length of a man's shoe: ______

27. The length of one big step: ______

Arrange each set of units in order from shortest to longest.

28. m, cm, mm, km ______, ______, ______, ______

29. ft, mi, in., yd ______, ______, ______, ______

Find each equivalent measure.

30. 10 cm = ______ mm

31. 2 m = ______ cm or ______ mm

32. 5 km = ______ m

33. 2.5 cm = ______ mm

34. 1.5 m = ______ cm or ______ mm

35. 7.5 km = ______ m

36. $\frac{1}{2}$ ft = ______ in.

37. 2 ft = ______ in.

38. 3 ft = ______ in.

39. 2 yd = ______ ft

40. 10 yd = ______ ft

41. 100 yd = ______ ft

FACTS PRACTICE TEST

L Liquid Measurement

Name ____________________

Time ____________________

Write the abbreviation for each of the following units.

Metric Units

1. liter ________

2. milliliter ________

U.S. Customary Units

3. ounce ________

4. pint ________

5. quart ________

6. gallon ________

Complete each unit conversion.

Metric Conversions

7. 1 liter = ________ milliliters

U.S. Customary Conversions

8. 1 cup = ________ ounces

9. 1 pint = ________ ounces

10. 1 quart = ________ pints

11. 1 gallon = ________ quarts

Conversions between systems

12. 1 liter ≈ ________ quarts

13. 1 gallon ≈ ________ liters

Complete each statement.

14. One fourth of a dollar is a quarter, and one fourth of a gallon is a ____________.

15. A two-liter bottle of soda is a little more than 2 ____________ or $\frac{1}{2}$ of a ____________.

16. "A pint's a pound the world around" means a ____________ of water weighs about 1 pound.

17. In the United States, gasoline is sold by the ____________. In other countries, gasoline is sold by the ____________. A gallon of gasoline is more than ________ liters but a little less than ________ liters.

18. 2 cups make a ____________.

19. 2 pints make a ____________.

20. 2 quarts make a ____________.

21. 2 half gallons make a ____________.

22. A gallon of milk will fill ________ cups.

23. If you drink 8 cups of water each day, you drink a ____________ of water.

Find each equivalent measure.

24. 2 liters = ________ milliliters

25. 2 liter ≈ ________ quarts

26. 3.78 liters = ________ milliliters

27. 0.5 liters = ________ milliliters

28. $\frac{1}{2}$ pint (1 cup) = ________ ounces

29. 1 quart (2 pints) = ________ ounces

30. $\frac{1}{2}$ gallon = ________ quarts

31. 1 gallon = ________ pints

32. 2 gallons = ________ quarts

FACTS PRACTICE TEST

M

24 Percent-Fraction-Decimal Equivalents

Name ______________________

Time ______________________

Write each percent as a reduced fraction and a decimal.

Percent	Fraction	Decimal
40%		
5%		
80%		
2%		
3%		
20%		
25%		
60%		
1%		
90%		
75%		
10%		

Percent	Fraction	Decimal
70%		
4%		
100%		
30%		
50%		
$12\frac{1}{2}$%		
$37\frac{1}{2}$%		
$62\frac{1}{2}$%		
$87\frac{1}{2}$%		
$33\frac{1}{3}$%		Rounds to 0.333
$66\frac{2}{3}$%		Rounds to 0.667
$16\frac{2}{3}$%		Rounds to 0.167

FACTS PRACTICE TEST

N

Measurement Facts

Name ____________________

Time ____________________

1. Draw a segment 1 cm long.
2. Draw a segment 1 in. long.
3. One inch is how many centimeters? ____________
4. Which is longer, 1 km or 1 mi? ____________
5. Which is longer, 1 km or $\frac{1}{2}$ mi? ____________
6. How many ounces are in a pound? ____________
7. How many pounds are in a ton? ____________
8. A dollar bill has a mass of about one ____________.
9. Your math book has a mass of about one ____________.
10. On earth a kilogram mass weighs about ________ pounds.
11. A metric ton is ________ kilograms.
12. On earth a metric ton weighs about ________ pounds.
13. The earth rotates on its axis once in a ____________.
14. The earth revolves around the sun once in a ____________.

Abbreviations:

15. milligram ________
16. gram ________
17. kilogram ________
18. ________ C
19. ounce ________
20. pound ________
21. ton ________
22. ________ F

Equivalents:

23. 1 gram = ________ milligrams
24. 1 kilogram = ________ grams
25. $\frac{1}{2}$ ton = ________ pounds
26. ________ days = a common year
27. ________ days = a leap year
28. ________ weeks ≈ a year
29. ________ years = a decade
30. ________ years = a century
31. ________ years = a millennium

How many days are in

32. Jan. ______
33. Feb. ______ or ______
34. Mar. ______
35. Apr. ______
36. May ______
37. June ______
38. July ______
39. Aug. ______
40. Sept. ______
41. Oct. ______
42. Nov. ______
43. Dec. ______

Write the indicated temperatures.

44. Water boils ______ °F
45. ______ °C
46. Normal body temperature ______ °F
47. ______ °C

Cool room temperature 68 °F

48. ______ °C
49. Water freezes ______ °F
50. ______ °C

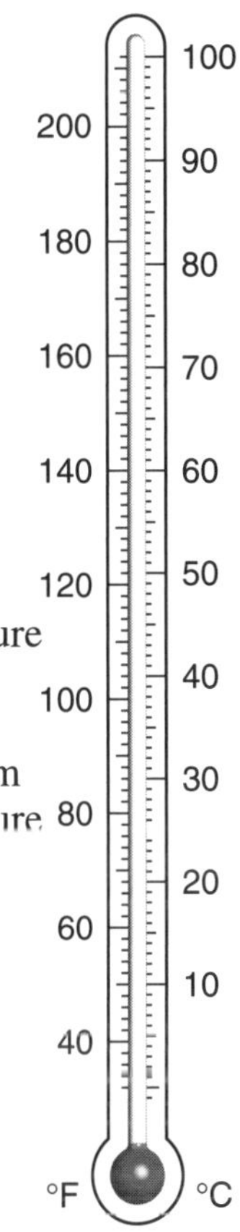

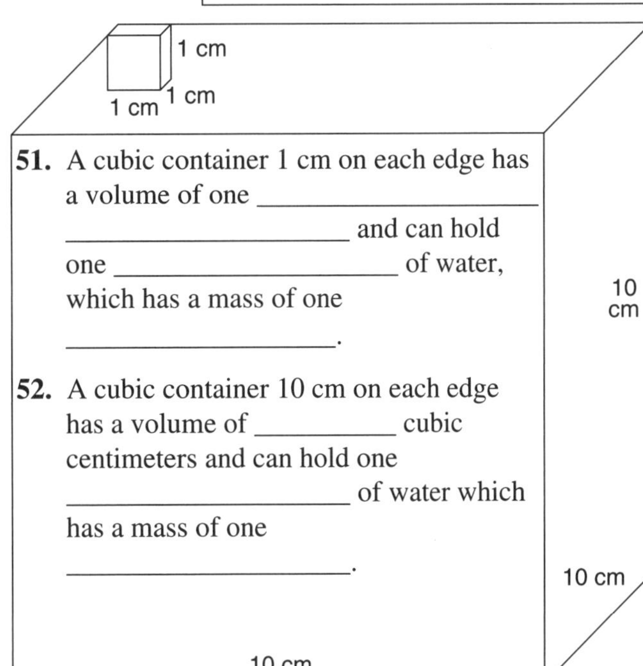

51. A cubic container 1 cm on each edge has a volume of one ____________ ____________ and can hold one ____________ of water, which has a mass of one ____________.

52. A cubic container 10 cm on each edge has a volume of ________ cubic centimeters and can hold one ____________ of water which has a mass of one ____________.

FACTS PRACTICE TEST

O Adding Integers

Name ______________________

Time ______________________

Simplify.

(+3) + (–4) =	(–2) + (+2) =	(–3) + (+4) =
(+2) + (+2) =	(–3) + (–4) =	(–2) + (–2) =
(+3) + (+4) =	(+2) + (–2) =	(–5) + (+2) =
(+5) + (–5) =	(–5) + (–2) =	(+5) + (–2) =
–6 + +4 =	–5 + –5 =	–6 + –4 =
+5 + –5 =	+6 + –4 =	+5 + +5 =
+10 + –5 =	–10 + +10 =	–10 + –5 =
+10 + –10 =	–10 + +5 =	–10 + –10 =
6 + (–3) =	5 + –8 =	–6 + 4 =
(–3) + –5 =	6 + (+3) =	–3 + 4 =
–6 + (–5) =	8 + (–2) =	–3 + (–4) =
–8 + (–8) =	8 + (–8) =	–8 + 8 =

FACTS PRACTICE TEST

P

Adding and Subtracting Integers

Name ______________________

Time ______________________

Simplify.

(–3) – (+3) =	(–4) + (+6) =	5 + –2 =
(–7) + (–1) =	10 – (–10) =	(–3) + (+3) =
10 – (–10) =	(–3) – (–3) =	–5 + 2 =
(–4) – (+6) =	(–7) – (–1) =	4 + (–6) =
10 + (–8) =	(+7) – (–1) =	(+3) – (–3) =
(+7) – (+1) =	5 – –2 =	(–4) + (–6) =
(+3) – (+3) =	–10 – 8 =	(–7) – (+1) =
10 + (–10) =	(–7) + (+1) =	–5 – 2 =
–5 + –2 =	(+3) + (+3) =	–10 + (–8) =
–10 – (–8) =	(–4) – (–6) =	(+7) + (+1) =
(–3) + (–3) =	10 – (–8) =	(+3) + (–3) =
4 – (–6) =	–5 – –2 =	(+7) + (–1) =

FACTS PRACTICE TEST

Q

Multiplying and Dividing Integers

Name ____________________

Time ____________________

Simplify.

$(-3)(-6) =$	$\frac{-8}{-4} =$	$(-2)(+5) =$	$\frac{+5}{-5} =$
$(+12) \div (+4) =$	$(-6)(+4) =$	$\frac{-15}{5} =$	$(+5)(+2) =$
$(-2)(+8) =$	$\frac{-5}{+5} =$	$(-6)(-4) =$	$\frac{-8}{+4} =$
$\frac{+5}{+5} =$	$(2)(+8) =$	$(+12) \div (-4) =$	$(3)(-6) =$
$(+3)(+6) =$	$\frac{-15}{-5} =$	$(-2)(-8) =$	$\frac{+8}{-4} =$
$(-12) \div (+4) =$	$(-5)(-2) =$	$\frac{+15}{+5} =$	$(+6)(-4) =$
$(-3)(6) =$	$\frac{+15}{-5} =$	$(+2)(-8) =$	$\frac{-5}{-5} =$
$\frac{+8}{+4} =$	$(+6)(+4) =$	$(-12) \div (-4) =$	$(-5)(+2) =$

FACTS PRACTICE TEST

R

90 Division Facts

Name ______________________

Time ______________________

Divide.

20 ÷ 4 =	21 ÷ 7 =	0 ÷ 2 =	27 ÷ 3 =	8 ÷ 1 =	54 ÷ 6 =
15 ÷ 5 =	6 ÷ 3 =	28 ÷ 4 =	18 ÷ 2 =	24 ÷ 6 =	9 ÷ 9 =
56 ÷ 8 =	0 ÷ 6 =	21 ÷ 3 =	1 ÷ 1 =	25 ÷ 5 =	12 ÷ 2 =
5 ÷ 1 =	45 ÷ 9 =	16 ÷ 4 =	30 ÷ 6 =	9 ÷ 3 =	14 ÷ 7 =
0 ÷ 8 =	6 ÷ 2 =	24 ÷ 8 =	10 ÷ 5 =	81 ÷ 9 =	24 ÷ 4 =
16 ÷ 2 =	30 ÷ 5 =	0 ÷ 1 =	28 ÷ 7 =	4 ÷ 4 =	40 ÷ 8 =
3 ÷ 3 =	18 ÷ 6 =	63 ÷ 9 =	40 ÷ 5 =	10 ÷ 2 =	36 ÷ 6 =
32 ÷ 8 =	12 ÷ 4 =	18 ÷ 3 =	35 ÷ 7 =	8 ÷ 8 =	2 ÷ 1 =
45 ÷ 5 =	7 ÷ 7 =	27 ÷ 9 =	9 ÷ 1 =	48 ÷ 6 =	0 ÷ 7 =
4 ÷ 1 =	0 ÷ 9 =	24 ÷ 3 =	32 ÷ 4 =	5 ÷ 5 =	72 ÷ 9 =
56 ÷ 7 =	15 ÷ 3 =	12 ÷ 6 =	8 ÷ 2 =	63 ÷ 7 =	0 ÷ 4 =
14 ÷ 2 =	42 ÷ 6 =	6 ÷ 1 =	16 ÷ 8 =	20 ÷ 5 =	49 ÷ 7 =
36 ÷ 4 =	64 ÷ 8 =	0 ÷ 3 =	54 ÷ 9 =	4 ÷ 2 =	48 ÷ 8 =
18 ÷ 9 =	3 ÷ 1 =	35 ÷ 5 =	8 ÷ 4 =	72 ÷ 8 =	6 ÷ 6 =
0 ÷ 5 =	42 ÷ 7 =	2 ÷ 2 =	36 ÷ 9 =	7 ÷ 1 =	12 ÷ 3 =

FACTS PRACTICE TEST

S

48 Divisions with Remainders

Name ____________________

Time ____________________

Divide.

$4\overline{)15}$	$9\overline{)14}$	$7\overline{)45}$	$3\overline{)16}$	$6\overline{)38}$	$2\overline{)7}$
$8\overline{)50}$	$5\overline{)28}$	$4\overline{)21}$	$6\overline{)15}$	$7\overline{)11}$	$8\overline{)20}$
$3\overline{)20}$	$7\overline{)32}$	$8\overline{)30}$	$2\overline{)15}$	$5\overline{)43}$	$6\overline{)35}$
$9\overline{)62}$	$4\overline{)10}$	$6\overline{)27}$	$9\overline{)21}$	$4\overline{)19}$	$3\overline{)25}$
$6\overline{)56}$	$2\overline{)17}$	$3\overline{)10}$	$5\overline{)8}$	$9\overline{)40}$	$7\overline{)30}$
$2\overline{)5}$	$8\overline{)25}$	$5\overline{)17}$	$7\overline{)17}$	$3\overline{)8}$	$4\overline{)9}$
$7\overline{)20}$	$6\overline{)10}$	$2\overline{)9}$	$4\overline{)30}$	$8\overline{)15}$	$9\overline{)29}$
$5\overline{)32}$	$3\overline{)14}$	$9\overline{)50}$	$8\overline{)65}$	$2\overline{)11}$	$5\overline{)19}$

FACTS PRACTICE TEST

T

60 Improper Fractions to Simplify

Name ____________________

Time ____________________

Write each improper fraction as a mixed number or whole number.

$\frac{15}{2} =$	$\frac{9}{8} =$	$\frac{10}{2} =$	$\frac{18}{6} =$	$\frac{8}{3} =$	$\frac{12}{4} =$
$\frac{10}{10} =$	$\frac{3}{2} =$	$\frac{11}{4} =$	$\frac{4}{3} =$	$\frac{12}{5} =$	$\frac{5}{4} =$
$\frac{12}{6} =$	$\frac{9}{3} =$	$\frac{5}{5} =$	$\frac{15}{4} =$	$\frac{6}{2} =$	$\frac{9}{9} =$
$\frac{3}{3} =$	$\frac{7}{4} =$	$\frac{21}{10} =$	$\frac{11}{2} =$	$\frac{7}{6} =$	$\frac{24}{8} =$
$\frac{11}{3} =$	$\frac{9}{5} =$	$\frac{4}{2} =$	$\frac{21}{8} =$	$\frac{6}{5} =$	$\frac{12}{3} =$
$\frac{7}{2} =$	$\frac{25}{6} =$	$\frac{10}{9} =$	$\frac{4}{4} =$	$\frac{12}{2} =$	$\frac{16}{15} =$
$\frac{10}{5} =$	$\frac{5}{2} =$	$\frac{7}{3} =$	$\frac{8}{4} =$	$\frac{8}{8} =$	$\frac{27}{10} =$
$\frac{16}{4} =$	$\frac{6}{6} =$	$\frac{25}{12} =$	$\frac{5}{3} =$	$\frac{7}{5} =$	$\frac{16}{9} =$
$\frac{15}{8} =$	$\frac{10}{3} =$	$\frac{33}{10} =$	$\frac{2}{2} =$	$\frac{35}{6} =$	$\frac{25}{8} =$
$\frac{6}{3} =$	$\frac{8}{5} =$	$\frac{9}{4} =$	$\frac{12}{12} =$	$\frac{25}{2} =$	$\frac{9}{2} =$

FACTS PRACTICE TEST

U

40 Fractions to Reduce

Name ______________________

Time ______________________

Reduce each fraction to lowest terms.

$\frac{2}{10} =$	$\frac{8}{16} =$	$\frac{2}{6} =$	$\frac{10}{100} =$	$\frac{6}{8} =$
$\frac{10}{15} =$	$\frac{5}{10} =$	$\frac{8}{12} =$	$\frac{9}{15} =$	$\frac{4}{16} =$
$\frac{2}{8} =$	$\frac{4}{10} =$	$\frac{15}{20} =$	$\frac{4}{8} =$	$\frac{4}{6} =$
$\frac{6}{15} =$	$\frac{4}{12} =$	$\frac{25}{100} =$	$\frac{10}{25} =$	$\frac{12}{20} =$
$\frac{20}{100} =$	$\frac{6}{9} =$	$\frac{2}{4} =$	$\frac{3}{12} =$	$\frac{3}{15} =$
$\frac{3}{9} =$	$\frac{2}{12} =$	$\frac{6}{10} =$	$\frac{12}{16} =$	$\frac{50}{100} =$
$\frac{9}{12} =$	$\frac{3}{6} =$	$\frac{5}{15} =$	$\frac{10}{12} =$	$\frac{8}{24} =$
$\frac{12}{15} =$	$\frac{8}{10} =$	$\frac{75}{100} =$	$\frac{6}{12} =$	$\frac{12}{24} =$

V

50 Fractions to Simplify

Name ____________________

Time ____________________

Simplify these fractions. Reduce the answers.

$\frac{16}{20}$ =	$\frac{6}{4}$ =	$\frac{4}{6}$ =	$\frac{10}{8}$ =	$\frac{3}{12}$ =
$\frac{12}{9}$ =	$\frac{2}{4}$ =	$\frac{12}{10}$ =	$\frac{12}{4}$ =	$\frac{12}{8}$ =
$\frac{8}{3}$ =	$\frac{8}{6}$ =	$\frac{4}{12}$ =	$\frac{10}{4}$ =	$\frac{4}{10}$ =
$\frac{20}{8}$ =	$\frac{4}{8}$ =	$\frac{20}{9}$ =	$\frac{24}{6}$ =	$\frac{9}{6}$ =
$\frac{15}{10}$ =	$\frac{5}{2}$ =	$\frac{12}{20}$ =	$\frac{15}{9}$ =	$\frac{8}{12}$ =
$\frac{4}{20}$ =	$\frac{8}{24}$ =	$\frac{10}{6}$ =	$\frac{3}{6}$ =	$\frac{16}{10}$ =
$\frac{2}{8}$ =	$\frac{20}{6}$ =	$\frac{6}{3}$ =	$\frac{25}{12}$ =	$\frac{9}{12}$ =
$\frac{10}{2}$ =	$\frac{8}{8}$ =	$\frac{50}{100}$ =	$\frac{6}{12}$ =	$\frac{15}{6}$ =
$\frac{10}{3}$ =	$\frac{10}{20}$ =	$\frac{24}{9}$ =	$\frac{6}{8}$ =	$\frac{16}{5}$ =
$\frac{5}{10}$ =	$\frac{14}{8}$ =	$\frac{15}{2}$ =	$\frac{21}{6}$ =	$\frac{16}{24}$ =

W

30 Percents to Write as Fractions

Name ______________________

Time ______________________

Write each percent as a reduced fraction.

1% =	20% =	55% =	90% =	75% =
99% =	5% =	95% =	80% =	12% =
70% =	65% =	50% =	2% =	48% =
24% =	25% =	98% =	40% =	15% =
60% =	30% =	4% =	35% =	36% =
45% =	8% =	10% =	21% =	85% =

X

24 Fraction-Decimal-Percent Equivalents

Name ________________

Time ________________

Write each fraction as a decimal and a percent.

Fraction	Decimal	Percent
$\frac{1}{100}$		
$\frac{1}{2}$		
$\frac{1}{20}$		
$\frac{1}{4}$		
$\frac{3}{4}$		
$\frac{1}{5}$		
$\frac{1}{50}$		
$\frac{1}{10}$		
$\frac{3}{10}$		
$\frac{2}{5}$		
$\frac{7}{10}$		
$\frac{3}{5}$		

Fraction	Decimal	Percent
$\frac{4}{5}$		
$\frac{9}{10}$		
$\frac{1}{3}$	Rounds to 0.333	
$\frac{2}{3}$	Rounds to 0.667	
$\frac{1}{8}$		
$\frac{7}{8}$		
$\frac{1}{25}$		
$\frac{1}{6}$	Rounds to 0.167	
$\frac{1}{9}$	Rounds to 0.111	
$\frac{3}{8}$		
$\frac{5}{8}$		
$\frac{5}{6}$	Rounds to 0.833	

Course Assessments

Pretest and Posttest

A fifty-problem, multiple-choice test to be given at the beginning and end of the school year to assess the progress students make

Final Exam

A fifty-problem, multiple-choice, cumulative test to be given at the completion of the course to assess students' comprehension of the topics in the course

Pretest and Posttest

Name ____________________________

1. $946 \div 22$ equals

A. 430 B. 403 C. 43 D. 4 R 66 E. None correct

2. $6.5 + 2.47 + 0.7$ equals

A. 3.19 B. 9.67 C. 9.04 D. 8.977 E. None correct

3. $23.45 - 1.2$ equals

A. 23.33 B. 24.65 C. 22.65 D. 22.25 E. None correct

4. 75×608 equals

A. 45,600 B. 5,100 C. 56,350 D. 52,460 E. None correct

5. $7.20 ÷ 20 equals

A. $3.60 B. $36.00 C. $144.00 D. 36¢ E. None correct

6. $3\frac{2}{3} + 2\frac{1}{3}$ equals

A. $5\frac{1}{2}$ B. 6 C. $1\frac{1}{3}$ D. $6\frac{2}{9}$ E. None correct

7. $4\frac{3}{4} - 2\frac{1}{4}$ equals

A. $2\frac{1}{2}$ B. $2\frac{1}{4}$ C. $1\frac{1}{2}$ D. 7 E. None correct

8. $\frac{3}{5} \times 10$ equals

A. 2 B. 4 C. 6 D. 8 E. None correct

9. $\frac{3}{4} \div \frac{1}{2}$ equals

A. $\frac{3}{8}$ B. $1\frac{1}{2}$ C. $\frac{2}{3}$ D. $2\frac{2}{3}$ E. None correct

10. $\frac{8}{12}$ reduces to

A. $1\frac{1}{2}$ B. $\frac{3}{4}$ C. $\frac{2}{3}$ D. $\frac{1}{4}$ E. None correct

11. Which digit in 125.6 is in the tenths place?

A. 1 B. 2 C. 5 D. 6 E. None correct

12. Which shows two million, five hundred thousand?

A. 2,500,000 B. 2,001,500 C. 2,501,000 D. 2,000,500 E. None correct

13. A segment 2 cm long is how many millimeters long?

A. 0.2 mm B. 2 mm C. 20 mm D. 200 mm E. None correct

14. Which is the best estimate for the width of a classroom door?

A. 1 kilometer B. 1 meter C. 1 centimeter D. 1 millimeter

15. Which is not equivalent to $\frac{1}{2}$?

A. $\frac{2}{4}$ B. 0.5 C. 50% D. 0.2

16. 12^2 equals

A. 14 B. 24 C. 122 D. 144 E. None correct

17. $\sqrt{100}$ equals

A. 5^2 B. 10^2 C. 10 D. 50 E. None correct

18. Which shows ten and three tenths?

A. 10.3 B. 10.03 C. 0.103 D. 103.0 E. None correct

19. Estimate the product of 692 and 412 by rounding each number to the nearest hundred before multiplying.

A. 240,000 B. 280,000 C. 28,000 D. 24,000

20. Which group of numbers is in order from least to greatest?

A. 0.01, 0.1, 1.0 B. 0.1, 0.01, 1.0 C. 1.0, 0.1, 0.01 D. 0.1, 1.0, 0.01 E. None correct

21. Which equals $\frac{1}{3}$?

A. $\frac{2}{5}$ B. $\frac{3}{6}$ C. $\frac{3}{9}$ D. $\frac{6}{9}$ E. None correct

22. Mike can ride 15 miles in one hour. At that rate, how far can he ride in 3 hours?

A. 5 miles B. 18 miles C. 30 miles D. 45 miles E. None correct

23. In three classrooms there were 22, 22, and 25 students. What was the average number of students in the three classrooms?

A. 22 students B. 23 students C. 24 students D. 25 students E. None correct

24. Kobe's test scores were 85, 80, 80, 85, 80, 90, and 95. What is the mode of the scores?

A. 80 B. 85 C. 15 D. 82.5 E. None correct

25. If the arrow is spun, what is the probability it will stop in region *A*?

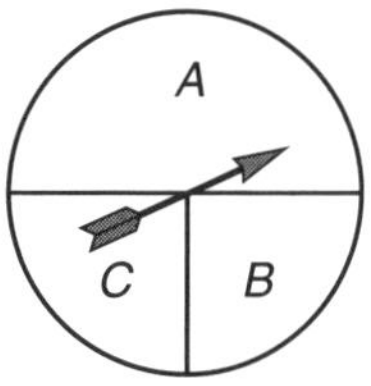

A. $\frac{1}{3}$ B. $\frac{1}{2}$ C. $\frac{1}{4}$ D. $\frac{2}{3}$ E. None correct

26. The meal cost $5.87. Tax was 49¢. What was the total including tax?

A. $5.38 B. $10.77 C. $6.36 D. $54.87 E. None correct

27. The newborn baby weighed 8 pounds. One pound is 16 ounces. How many ounces is 8 pounds?

A. 128 ounces B. 64 ounces C. 256 ounces D. 2 ounces E. None correct

28. Shirts regularly priced at $27.50 were on sale for half price. What was the sale price of a shirt?

A. $13.25 B. $13.75 C. $14.25 D. $27.00 E. None correct

29. Ling correctly answered 16 of the 20 questions. What fraction of the questions did Ling answer correctly?

A. $\frac{1}{5}$ B. $\frac{3}{4}$ C. $\frac{1}{4}$ D. $\frac{4}{5}$ E. None correct

30. At 60 miles per hour, how far can the train travel in 5 hours?

A. 300 miles B. 120 miles C. 3000 miles D. 12 miles E. None correct

31. Which triangle does not appear to be a right triangle?

A. 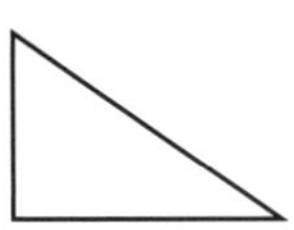 B. 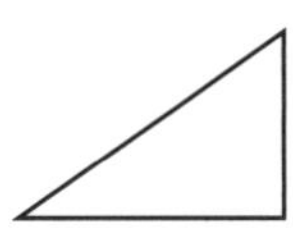 C. 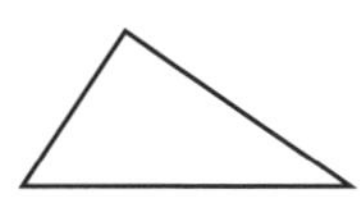 D.  E. None correct

32. Which angle appears to be an acute angle?

A. 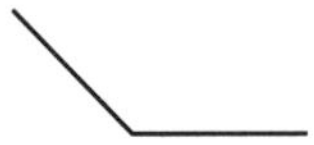 B.  C.  D.  E. None correct

33. What is the perimeter of the rectangle?

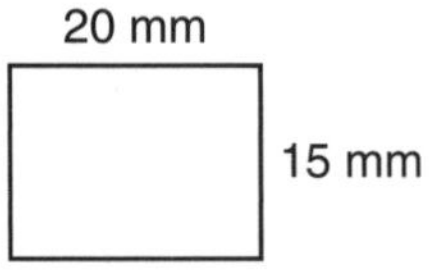

A. 35 mm B. 70 mm C. 300 mm D. 75 mm

34. The diameter of a wheel is 40 inches. What is the radius of the wheel?

A. 10 inches B. 20 inches C. 80 inches D. 120 inches E. None correct

35. What is the area of the rectangle?

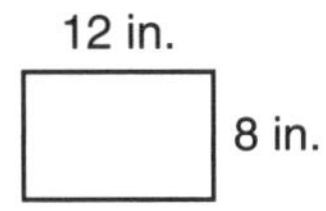

A. 20 sq. in. B. 40 sq. in. C. 96 sq. in. D. 106 sq. in.

36. What fraction of the circle is shaded?

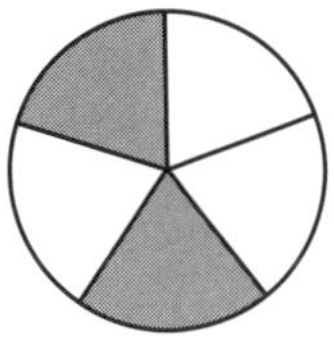

A. $\frac{2}{5}$ B. $\frac{2}{3}$ C. $\frac{3}{5}$ D. $\frac{1}{5}$ E. None correct

37. A cube has how many faces?

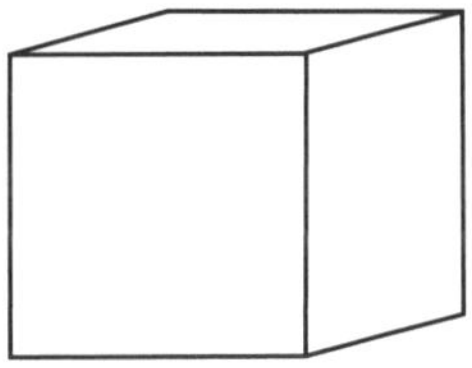

A. 12 B. 3 C. 6 D. 8

38. Which triangle below appears to be congruent to this triangle?

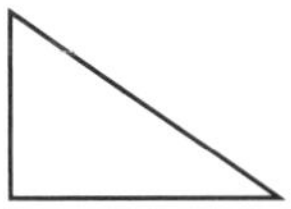

A. 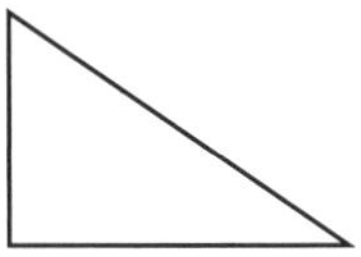
B. 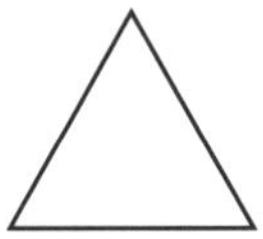
C. 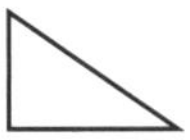
D. 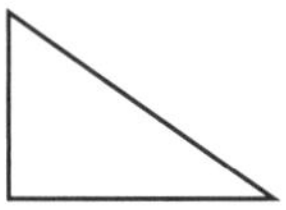

39. Which pair of lines appears to be parallel?

A. 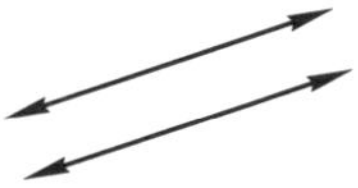
B. 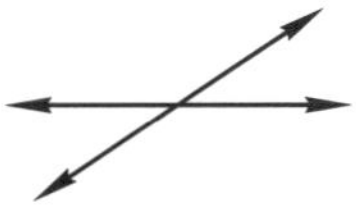
C. 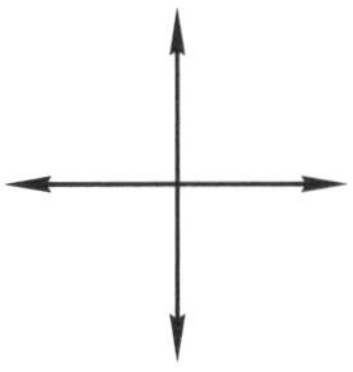
D. 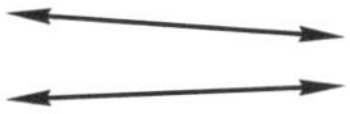

40. The area of the square is 1 square yard. The area is also how many square feet?

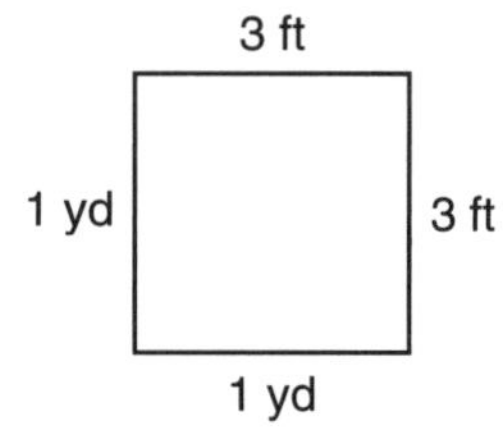

A. 3 sq. ft B. 6 sq. ft C. 9 sq. ft D. 12 sq. ft

41. Which comparison is true?

A. $-5 < -7$ B. $-2 < -1$ C. $-3 > 0$ D. $2 < -3$ E. None correct

42. If $n - 12 = 15$, then n equals

A. 27 B. 18 C. 17 D. 3 E. None correct

43. $5\frac{2}{3} - \left(4 - 1\frac{1}{3}\right)$ equals

A. $\frac{1}{3}$ B. $3\frac{2}{3}$ C. 4 D. 3 E. None correct

44. Which of these numbers is not a factor of 12?

A. 2 B. 3 C. 4 D. 5 E. None correct

45. Which does not equal 6×12?

A. $(6 \times 10) + (6 \times 2)$ B. $(6 \times 6) + (6 \times 6)$
C. $(6 \times 3) + (6 \times 4)$ D. $(3 \times 12) + (3 \times 12)$

46. If $3n = 36$, then n equals

A. 6 B. 9 C. 12 D. 33 E. None correct

47. Which of the following shows that the product of $\frac{1}{2}$ and $\frac{1}{2}$ is less than the sum of $\frac{1}{2}$ and $\frac{1}{2}$?

A. $\frac{1}{2} \times \frac{1}{2} < \frac{1}{2} + \frac{1}{2}$

B. $\frac{1}{2} \times \frac{1}{2} > \frac{1}{2} + \frac{1}{2}$

C. $\frac{1}{2} - \frac{1}{2} < \frac{1}{2} + \frac{1}{2}$

D. $\frac{1}{2} \div \frac{1}{2} < \frac{1}{2} + \frac{1}{2}$

48. If $2x$ equals 12, then $2 + x$ equals

A. 6 B. 8 C. 4 D. 14 E. None correct

49. Point *A* could represent which of these numbers?

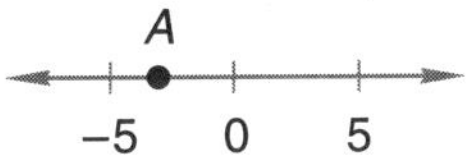

A. $\frac{1}{2}$ B. 3 C. –3 D. –8

50. Which point has the coordinate (–3, 2)?

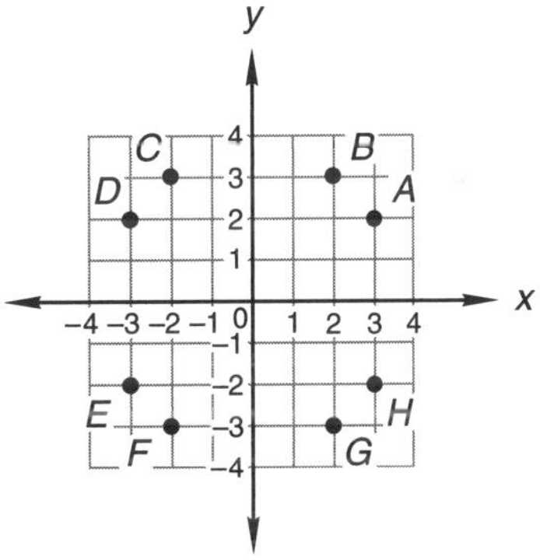

A. Point *A*

B. Point *B*

C. Point *C*

D. Point *D*

E. None correct

Final Exam

Name ______________________________

1. (25) $7610 \div 25$ equals

A. $34\frac{2}{5}$ B. $340\frac{2}{5}$ C. $304\frac{2}{5}$ D. $300\frac{2}{5}$ E. None correct

2. (38) $8.75 + 6 + 4.5$ equals

A. 20.25 B. 9.26 C. 13.31 D. 19.25 E. None correct

3. (40) $1.8 - 0.25$ equals

A. 1.55 B. 1.65 C. 1.3 D. 0.7 E. None correct

4. (39) 0.15×4.2 equals

A. 0.63 B. 6.3 C. 630 D. 0.063 E. None correct

5. (49) $3.64 \div 0.7$ equals

A. 0.52 B. 5.2 C. 52 D. 520 E. None correct

6. (59) $1\frac{2}{3} + 2\frac{5}{6}$ equals

A. $3\frac{7}{9}$ B. $4\frac{1}{2}$ C. $3\frac{1}{2}$ D. $4\frac{1}{6}$ E. None correct

7. (63) $3\frac{2}{5} - 1\frac{1}{2}$ equals

A. $2\frac{1}{3}$ B. $2\frac{1}{10}$ C. $1\frac{9}{10}$ D. $4\frac{9}{10}$ E. None correct

8. (66) $\frac{4}{5} \times 3\frac{1}{3}$ equals

A. $3\frac{4}{15}$ B. $1\frac{13}{15}$ C. $2\frac{2}{3}$ D. $\frac{3}{25}$ E. None correct

9. (68) $2\frac{2}{5} \div 1\frac{1}{2}$ equals

A. $1\frac{3}{5}$ B. $2\frac{1}{2}$ C. $3\frac{3}{5}$ D. $\frac{5}{8}$ E. None correct

10. (29) $\frac{60}{84}$, reduced to lowest terms, equals

A. $\frac{15}{21}$ B. $\frac{30}{42}$ C. $\frac{5}{7}$ D. $\frac{1}{2}$ E. None correct

11. (12) Which digit in 64,327.891 is in the thousands place?

A. 6 B. 4 C. 9 D. 1 E. None correct

12. (113) $2\frac{1}{2}$ million is

A. 2,500,000 B. 2,500,000,000 C. $2{,}000{,}000\frac{1}{2}$ D. 2,200,000 E. None correct

13. (95) Which of these equals 1.2 m?

A. 1200 cm B. 0.12 km C. 1200 mm D. 12 m E. None correct

14. (7) Which is the most reasonable measure for the length of a bicycle?

A. 2 mm B. 2 m C. 2 cm D. 2 km E. None correct

15. (99) 40% is NOT equivalent to

A. 0.4 B. $\frac{2}{5}$ C. $\frac{40}{100}$ D. 0.04 E. None correct

16. (92) $10^3 \cdot 10^4$ equals

A. 10^{12} B. 10^7 C. 1 million D. 1200 E. None correct

17. (89) $\sqrt{199}$ is between

A. 99 and 100 B. 9 and 10 C. 14 and 15 D. 19 and 20 E. None correct

18. (35) Twenty-five thousandths may be written

A. 0.25 B. 25,000 C. 20,005 D. 20.005 E. None correct

19. (51) Estimate the sum of 17.36 and 8.7 by rounding to the nearest whole number before adding.

A. 9 B. 10 C. 25 D. 26 E. None correct

20. (99) Which set of numbers is arranged from least to greatest?

A. $33\%, \frac{1}{3}, 0.3$ B. $0.3, \frac{1}{3}, 33\%$ C. $0.3, 33\%, \frac{1}{3}$ D. $\frac{1}{3}, 0.3, 33\frac{1}{3}\%$ E. None correct

21. (85) What number solves this proportion? $\frac{8}{12} = \frac{12}{n}$

A. 8 B. 16 C. 18 D. 24 E. None correct

22. (101) The ratio of boys to girls in the class was 2 to 3. If there were 18 girls, how many students were in the class?

A. 12 B. 24 C. 27 D. 30 E. None correct

23. (18) The average age of three people is 25. If two of the people are 22, how old is the third person?

A. 23 B. 25 C. 28 D. 31 E. None correct

24. (Inv. 5) Ten children measured and recorded how high they could jump. The results they recorded, in inches, were as follows:

8, 9, 8, 6, 8, 12, 9, 10, 11, 16

What is the median of these measures?

A. 8 B. 9 C. 9.5 D. 10 E. None correct

25. *(58)* One white marble, two red marbles, and three blue marbles are in a bag. If one marble is taken from the bag, what is the probability that it will be blue?

A. $\frac{1}{6}$　　B. $\frac{1}{3}$　　C. $\frac{1}{2}$　　D. $\frac{3}{5}$　　E. None correct

26. *(41)* If the price of an item is \$2.89 and sales tax is 8%, what is the total cost of the item including sales tax?

A. \$2.97　　B. \$3.12　　C. \$3.17　　D. \$23.12　　E. None correct

27. *(15)* What is the price per ounce of a 32-ounce box of cereal priced at \$4.16?

A. 11¢/oz　　B. 12¢/oz　　C. 13¢/oz　　D. 16¢/oz　　E. None correct

28. *(41)* A shirt regularly priced at \$36.00 was on sale for 25% off. What was the sale price?

A. \$9.00　　B. \$24.00　　C. \$27.00　　D. \$48.00　　E. None correct

29. *(75)* Emma correctly answered 21 of 25 questions. What percent of the questions did Emma answer correctly?

A. 21%　　B. 84%　　C. 16%　　D. 96%　　E. None correct

30. *(15)* At 400 miles per hour, how far can an airplane fly in $2\frac{1}{2}$ hours?

A. 600 miles　　B. 800 miles　　C. 650 miles　　D. 1000 miles　　E. None correct

31. *(93)* Which of these triangles appears to be equilateral?

A. 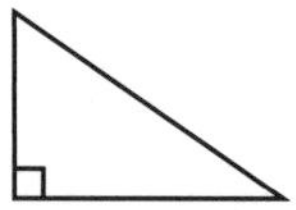　　B. 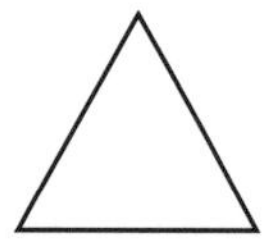　　C. 　　D. 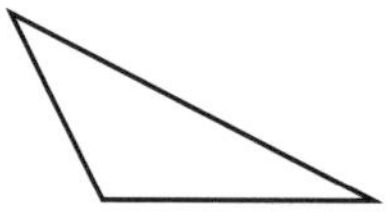　　E. None correct

32. *(98)* If $\angle x$ measures 38°, what is the measure of $\angle y$?

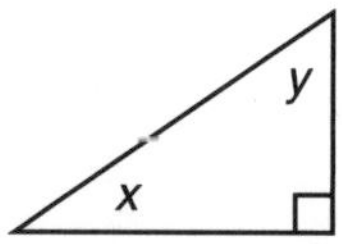

A. 52°　　B. 62°　　C. 38°　　D. 142°　　E. None correct

　　

33. (103) What is the perimeter of this figure? (All angles are right angles.)

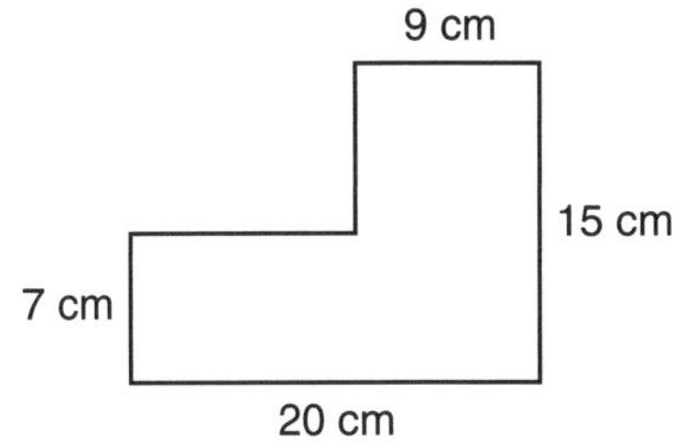

A. 212 cm B. 67 cm C. 51 cm D. 70 cm E. None correct

34. (47) If the radius of a bicycle tire is 10 inches, what is its circumference? (Use 3.14 for π.)

A. 31.4 in. B. 62.8 in. C. 314 in. D. 20 in. E. None correct

35. (79) What is the area of this triangle?

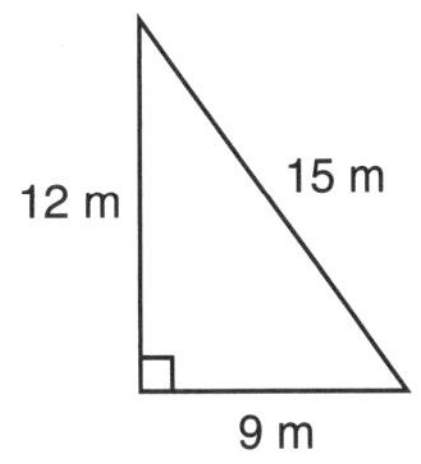

A. 108 m^2 B. 54 m^2 C. 36 m^2 D. 21 m^2 E. None correct

36. (86) The diameter of a circular braided rug is 8 feet. Which choice best describes the approximate area covered by the rug?

A. 50 ft^2 B. 25 ft^2 C. 64 ft^2 D. 16 ft^2

37. (82) What is the volume of this small box?

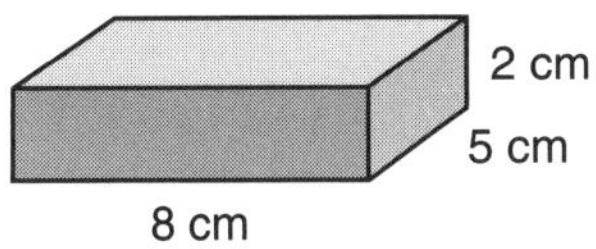

A. 80 cm^3 B. 40 cm^3 C. 30 cm^3 D. 15 cm^3 E. None correct

38. Which triangle below appears to be similar to ΔXYZ?
(109)

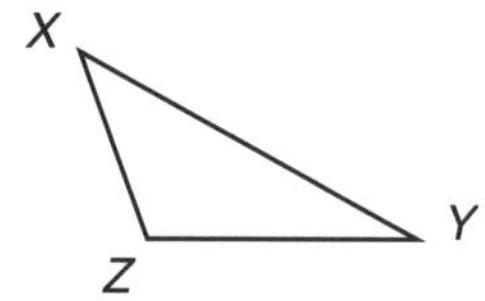

A. 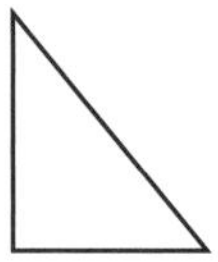

B. 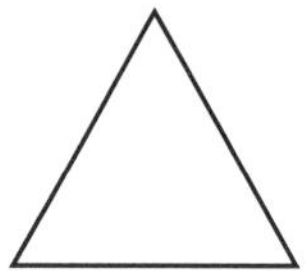

C. 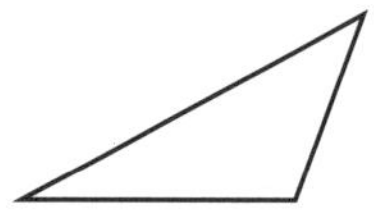

D. 

39. In rectangle *QRST,* which segment is parallel to $\overline{QR}$?
(28)

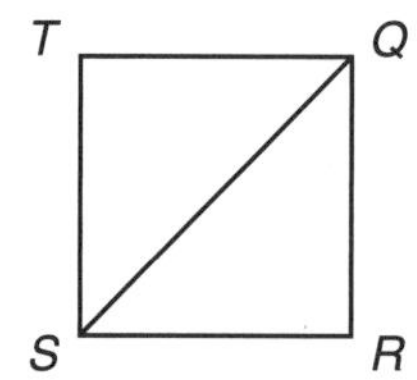

A. $\overline{QS}$ B. $\overline{QT}$ C. $\overline{RS}$ D. $\overline{TS}$ E. None correct

40. If $3^2 + 4^2 = c^2$, then c equals
(38, 92, 106)

A. 5 B. 7 C. 25 D. 49 E. None correct

41. $(-6) + (-3) + (+2)$ equals
(100)

A. –5 B. –7 C. –1 D. –11 E. None correct

42. If a is 3 and c is 12, then $4ac$ equals
(96)

A. 19 B. 36 C. 48 D. 144 E. None correct

43. $3[16 - 2(5 - 3)]$ equals
(5)

A. 9 B. 35 C. 36 D. 44 E. None correct

44. Which choice shows the prime factorization of 24?
(65, 73)

A. $2^3 \cdot 3$ B. $2 \cdot 12$ C. $3 \cdot 8$ D. $6 \cdot 4$ E. None correct

45. $6(25)$ equals
(2)

A. $6 \cdot 20 \cdot 5$ B. $6 \cdot 24 + 5$ C. $6 \cdot 20 + 5$ D. $6 \cdot 20 + 6 \cdot 5$

46. If $3m - 1 = 35$, then m equals
(106)

A. 6 B. 12 C. 34 D. 36 E. None correct

47. Which of these inequalities shows that the square root of 9 is less than 2 squared?
(9, 38)

A. $\sqrt{9} < 2^2$ B. $\sqrt{9} > 2^2$ C. $9^2 < \sqrt{2}$ D. $9^2 > \sqrt{2}$ E. None correct

48. $(-10)(-10)$ equals
(112)

A. –100 B. 100 C. –20 D. 0 E. None correct

49. If $x \geq -2$, which of these numbers cannot be used in place of x to make the inequality true?
(96)

A. $\frac{1}{2}$ B. –1 C. 0 D. –3 E. 5

50. The coordinates of three vertices of a rectangle are (–2, 3), (4, 3), and (4, –3). The coordinates of the fourth vertex are
(Inv. 7)

A. (–2, –3) B. (4, –2) C. (–2, 4) D. (–3, 3) E. None correct

Tests

Tests should be given after every fifth lesson, beginning after Lesson 10. The testing schedule is explained in greater detail on the back of this page.

Two forms of each test are included, providing the following options:

1. Use one form as an original test and the other as a makeup test.

2. Use both forms on the test day to discourage copying.

3. Use one form as an in-class practice (cooperative work acceptable) and the other as the test.

Additional test forms may be created using the *Saxon Math 7/6 Test and Practice Generator.*

On test days we recommend three activities:

1. Administer the Facts Practice Test indicated at the top of each test.

2. Administer the test.

3. Conduct the Test-Day Activity suggested in the Teacher's Manual.

Optional student answer forms are located in the Answer Forms section of this book. Each form provides a structure for students to show their work.

To grade the tests, refer to the Test Answers or Test Solutions in the back of this book.

Testing Schedule

Test to be administered	Covers material through	Give after teaching
Test 1	Lesson 5	Lesson 10
Test 2	Lesson 10	Lesson 15
Test 3	Lesson 15	Lesson 20
Test 4	Lesson 20	Lesson 25
Test 5	Lesson 25	Lesson 30
Test 6	Lesson 30	Lesson 35
Test 7	Lesson 35	Lesson 40
Test 8	Lesson 40	Lesson 45
Test 9	Lesson 45	Lesson 50
Test 10	Lesson 50	Lesson 55
Test 11	Lesson 55	Lesson 60
Test 12	Lesson 60	Lesson 65
Test 13	Lesson 65	Lesson 70
Test 14	Lesson 70	Lesson 75
Test 15	Lesson 75	Lesson 80
Test 16	Lesson 80	Lesson 85
Test 17	Lesson 85	Lesson 90
Test 18	Lesson 90	Lesson 95
Test 19	Lesson 95	Lesson 100
Test 20	Lesson 100	Lesson 105
Test 21	Lesson 105	Lesson 110
Test 22	Lesson 110	Lesson 115
Test 23	Lesson 115	Lesson 120

TEST

1A

Also take Facts Practice Test A (64 Addition Facts).

Name ______________________________

1. (1) Use the numbers 6, 12, and 18 to make two addition facts and two subtraction facts.

2. (2) Use the numbers 5, 15, and 75 to make two multiplication facts and two division facts.

3. (1) What is the sum of 3675, 285, and 1308?

4. (1) Patrick paid $5 for a $2.85 lunch. How much money should he get back?

5. (2) Jill bought 4 tickets for $1.75 each. What was the total cost of the tickets?

6. (2) Four quarters are worth $1.00. A roll of quarters is worth $10. How many quarters are in a roll?

7. (2) If 240 students are divided into groups of 30, how many groups will be formed?

8. (1) Alberto had 3 dozen pencils. Then he lost 1 pencil. Now how many pencils does Alberto have?

9. (1) $\$3.50 + \$0.95 + \$12.00$

10. (2) $28 + 28 + 28 + 28 + 28 + 28$

11. (1) $3826 - 382$

12. (2) $563 \cdot 48$

13. (2) $\begin{array}{r} 408 \\ \times\ \ 67 \\ \hline \end{array}$

14. (2) $\dfrac{9616}{8}$

15. (2) $24\overline{)3000}$

16. (5) $20 - (10 - 5)$

17. (5) $36 \div (6 \div 2)$

Find each missing number. Check your work.

18. (3) $47 + m = 100$

19. (3) $1000 - n = 258$

20. (4) $5x = 85$

TEST

1B

Also take Facts Practice Test A (64 Addition Facts).

Name ______________________

1. (1) Use the numbers 5, 11, and 16 to make two addition facts and two subtraction facts.

2. (2) Use the numbers 4, 16, and 64 to make two multiplication facts and two division facts.

3. (1) What is the total of \$1.25, \$0.98, and \$10?

4. (1) Terrell gave \$10 for a \$3.49 book. How much change should he get back?

5. (2) What is the total cost of one dozen pens if each pen costs \$0.29?

6. (2) Three quarters is the same amount of money as how many nickels?

7. (2) If 60 players are separated into 5 equal teams, how many players will be on each team?

8. (1) How many days is 4 weeks plus 3 days?

9. (1) $5986 + 367 + 1049$

10. (2) $365 + 365 + 365 + 365$

11. (1) $4216 - 377$

12. (2) $575 \cdot 16$

13. (2) $\begin{array}{r} 306 \\ \times\ \ 40 \\ \hline \end{array}$

14. (2) $\dfrac{7252}{7}$

15. (2) $25\overline{)3600}$

16. (5) $56 - (23 + 12)$

17. (5) $48 \div (3 \times 4)$

Find each missing number. Check your work.

18. (3) $67 + m = 114$

19. (3) $3000 - n = 1362$

20. (4) $4x = 72$

TEST

2A

Also take Facts Practice Test C (100 Subtraction Facts).

Name ______________________________

1. (1) Mark bought a pen for \$0.49, a box of pencils for \$1.29, and a pad of paper for \$0.98. How much did he spend in all?

2. (2) The coach bought 10 ice cream cones for \$0.35 each. How much did he spend?

3. (1) $\begin{array}{r} 3496 \\ 788 \\ +\ 1409 \\ \hline \end{array}$

4. (1) $\begin{array}{r} 4806 \\ -\ 1837 \\ \hline \end{array}$

5. (2) $\begin{array}{r} 47 \\ \times\ 36 \\ \hline \end{array}$

6. (2) $\dfrac{420}{15}$

7. (1) \$5 + \$3.49 + \$0.28

8. (5) $24 - (12 - 6)$

9. (2) $8\overline{)4048}$

Find each missing number:

10. (3) $67 + n = 130$

11. (4) $9m = 306$

12. (6) What fraction of this circle is shaded?

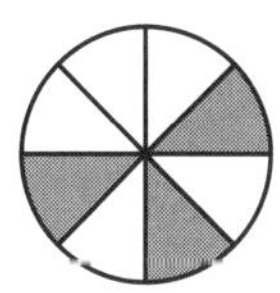

13. (6) What number is $\frac{1}{2}$ of 80?

14. (7) How long is this line segment?

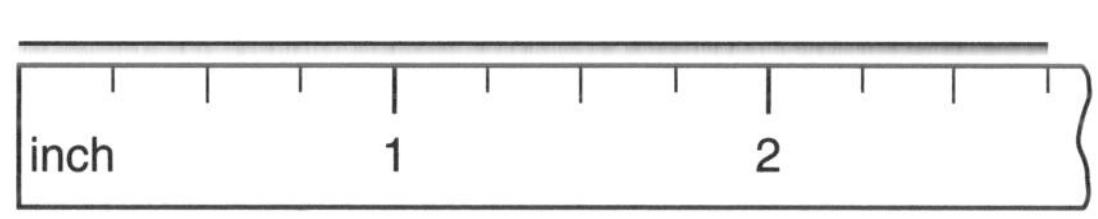

15. (9) Compare: $6 \times 7 \bigcirc 50 - 10$

16. (8) What is the perimeter of this rectangle?

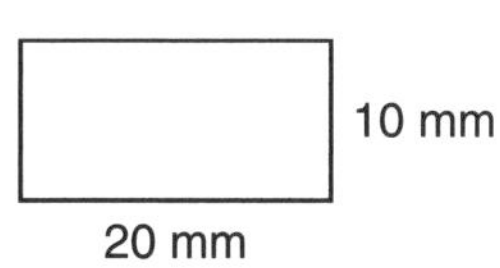

17. (10) What number is next in this sequence?

2, 6, 10, 14, _____

18. (10) Which of the following numbers is odd?

A. 888 B. 1234 C. 4321

19. (2) Here are three ways to show "15 divided by 5."

$5\overline{)15}$ $15 \div 5$ $\dfrac{15}{5}$

Write three ways to show "20 divided by 4."

20. (1) Arrange the numbers 10, 20, and 30 to form two addition facts and two subtraction facts.

TEST

2B

Also take Facts Practice Test C (100 Subtraction Facts).

Name ______________________

1. (1) Mai paid \$10 for a record that cost \$7.37. How much change should she get back?

2. (2) If two dozen cookies are divided evenly among 8 people, how many cookies will each person receive?

3. (1) $678 + 48 + 1386$

4. (1) $503 - 86$

5. (2) 34×36

6. (2) $3612 \div 6$

7. (1) \$6.25 + \$12 + \$0.59

8. (5) $18 - (9 + 6)$

9. (2) $25\overline{)800}$

Find each missing number:

10. (3) $46 + n = 81$

11. (4) $6m = 306$

12. (6) What fraction of this rectangle is shaded?

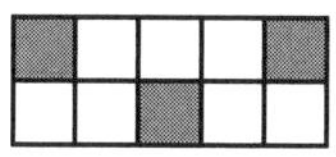

13. (6) What number is $\frac{1}{2}$ of 36?

14. (7) How long is this line segment?

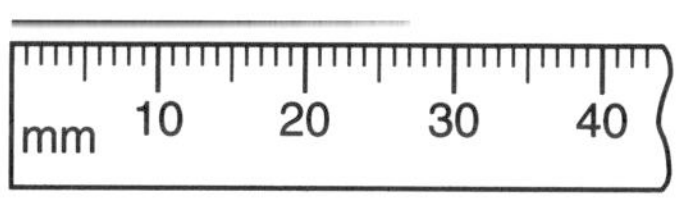

15. (9) Compare: $12 \times 7 \bigcirc 7 \times 12$

16. (8) What is the perimeter of this square?

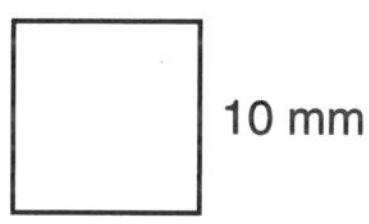

17. (10) What number is next in this sequence?

1, 2, 4, 8, 16, _____

18. (10) Which of the following numbers is even?

A. 2345 B. 7777 C. 7654

19. (2) Here are three ways to show "15 divided by 3."

$3\overline{)15}$ $15 \div 3$ $\frac{15}{3}$

Write three ways to show "16 divided by 8."

20. (1) Arrange the numbers 11, 22, and 33 to form two addition facts and two subtraction facts.

TEST

3A

Also take Facts Practice Test D (64 Multiplication Facts).

Name ______________________

1. If 1000 pennies are put into rolls of 50 pennies, how many rolls will be made?
(15)

2. How many years were there from 1215 to 1776?
(13)

3. The zero holds what place in 908,321?
(12)

4. Write 10,000,000 in word form.
(12)

5. What is the quotient when the sum of 8 and 4 is divided by the difference of 8 and 4?
(12)

6. What is the product of five thousand twenty and eighty-six?
(12)

7. Jason paused on step 189 on his way to the top of the Statue of Liberty. If there were 354 steps in all, how many more steps did Jason have to climb?
(11)

8. What temperature is shown on this thermometer?
(10)

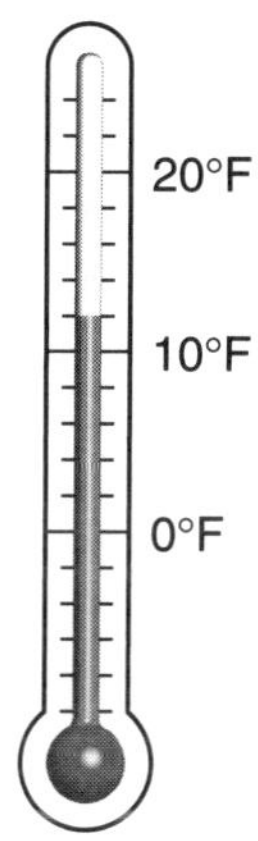

9. What fraction of this rectangle is shaded?
(6)

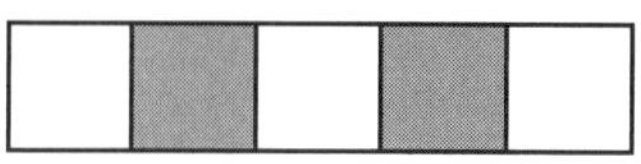

10. Compare: $-7 \bigcirc -4$
(14)

11. $5874 + 36{,}287 + 489$
(1)

12. $(48)(67)$
(2)

13. $\$1.25 \times 8$
(2)

14. $3542 \div 7$
(2)

15. $625 \div 25$
(2)

16. $24 \div (12 \div 4)$
(5)

Find each missing number:

17. $1000 - m = 408$
(3)

18. $9n = 63$
(4)

19. What is the perimeter of this square?
(8)

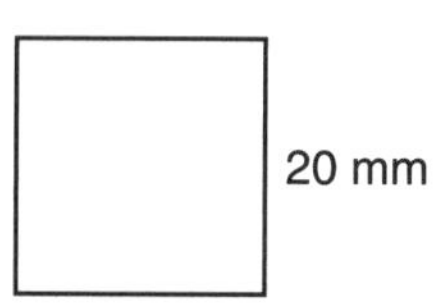

20. Use the numbers 8, 15, and 120 to make two multiplication facts and two division facts.
(2)

TEST

3B

Also take Facts Practice Test D (64 Multiplication Facts).

Name ______________________

1. (15) If 234 players are separated into 18 equal teams, how many players will be on each team?

2. (13) How many years were there from 1620 to 1776?

3. (12) What is the place value of the 6 in 123,456,789?

4. (12) Write 4,000,000,000 in word form.

5. (12) What is the difference when the sum of 4 and 5 is subtracted from the product of 4 and 5?

6. (12) When two hundred forty is divided by twelve, what is the quotient?

7. (11) Jasmine paused on step 178 on her way to the top of the Statue of Liberty. If there were 354 steps in all, how many more steps did Jasmine have to climb?

8. (10) What temperature is shown on this thermometer?

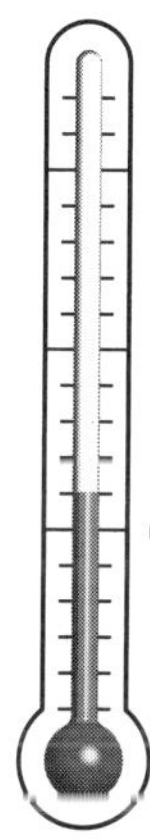

9. (6) What fraction of this circle is shaded?

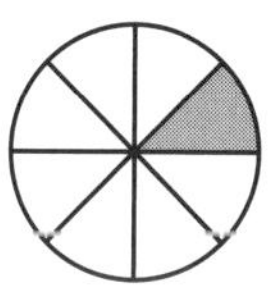

10. (14) Compare: $8 - 5 \bigcirc 5 - 8$

11. (1) $3.75 + $0.48 + $5

12. (2) (56)(83)

13. (2) $0.35 × 10

14. (2) 4824 ÷ 12

15. (2) $12.51 ÷ 3

16. (5) 24 ÷ (12 ÷ 6)

Find each missing number:

17. (3) $3000 - a = 1096$

18. (4) $9m = 54$

19. (8) What is the perimeter of this rectangle?

2 cm

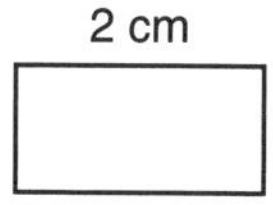

1 cm

20. (1) Use the numbers 8, 15, and 23 to make two addition facts and two subtraction facts.

TEST 4A

Also take Facts Practice Test D (64 Multiplication Facts).

Name ____________________

1. (12) When the sum of 9 and 3 is divided by the difference of 9 and 3, what is the quotient?

2. (12) Which digit is in the millions place in 123,456,789?

3. (11) Tom has read 96 pages of a 312-page book. How many pages does Tom need to read to finish the book?

4. (14) The morning temperature was –5°. By noon the temperature was 4°. How many degrees had the temperature risen?

Find each missing number:

5. (3) $Q - 40 = 25$

6. (4) $3n = 87$

7. (16) Round 37,542 to the nearest thousand.

8. (16) Estimate the product of 51 and 48.

9. (2) $\frac{123}{3}$

10. (18) What is the average of 3, 7, 9, 10, 10, and 9?

11. (18) What number is halfway between 24 and 98?

12. (19) List the whole-number factors of 18.

13. (15) The post office sold 39-cent commemorative stamps in sheets of 25 stamps. What was the price for the sheet of stamps?

14. (20) Find the greatest common factor (GCF) of 18 and 24.

Use the graph at right to answer questions 15 and 16.

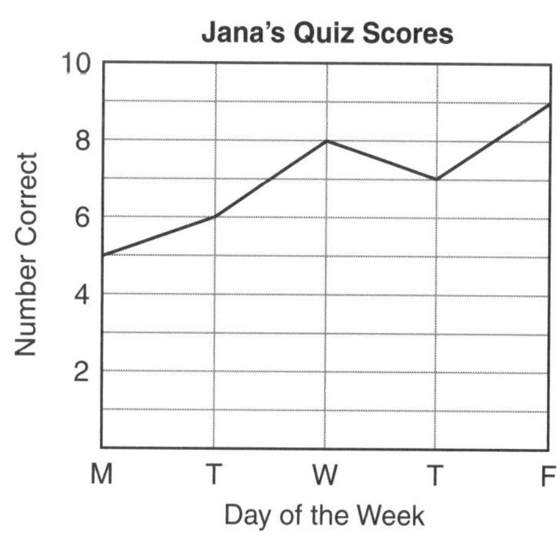

15. (18) On which day did Jana earn her highest score?

16. (18) How many answers did she have correct on Monday?

17. (6) What number is $\frac{1}{2}$ of \$3.50?

18. (8) What is the perimeter of this rectangle?

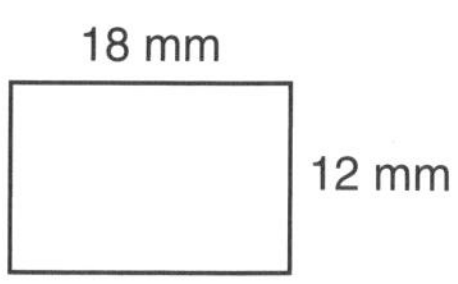

19. (10) What is the next number in this sequence?

15, 30, 45, 60, _____

20. (7) How long is this line segment?

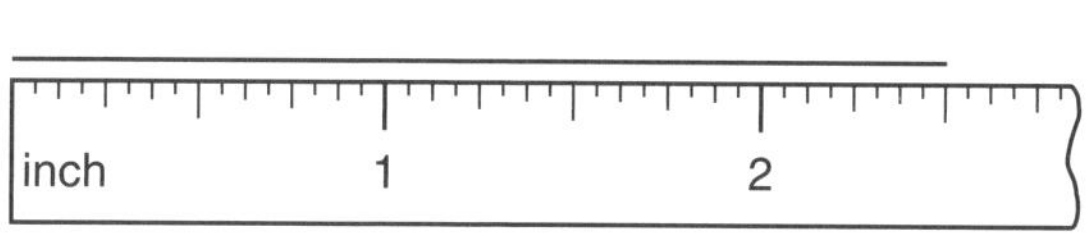

TEST

4B

Also take Facts Practice Test D (64 Multiplication Facts).

Name ______________________

1. (12) What is the difference when the sum of 3 and 4 is subtracted from the product of 4 and 3?

2. (12) Which digit is in the ten-thousands place in 987,654,321?

3. (11) Christina has read 94 pages of a 211-page book. How many pages does Christina need to read to finish the book?

4. (14) When Nia awoke, the temperature was –8°. By noon the temperature had risen to 10°. How many degrees had the temperature risen?

Find each missing number:

5. (3) $R - 56 = 62$

6. (4) $3m = 78$

7. (16) Round 4367 to the nearest hundred.

8. (16) Estimate the sum of 5280 and 3910 to the nearest thousand.

9. (2) $\frac{432}{6}$

10. (18) What is the average of 6, 10, 8, 7, and 9?

11. (18) What number is halfway between 17 and 83?

12. (19) List the whole-number factors of 21.

13. (15) The post office sold 37-cent commemorative stamps in sheets of 25 stamps. What was the price for the sheet of stamps?

14. (20) What is the greatest common factor (GCF) of 16 and 24?

Use the graph below to answer problems 15 and 16.

Points Distribution in Intramural Basketball Game

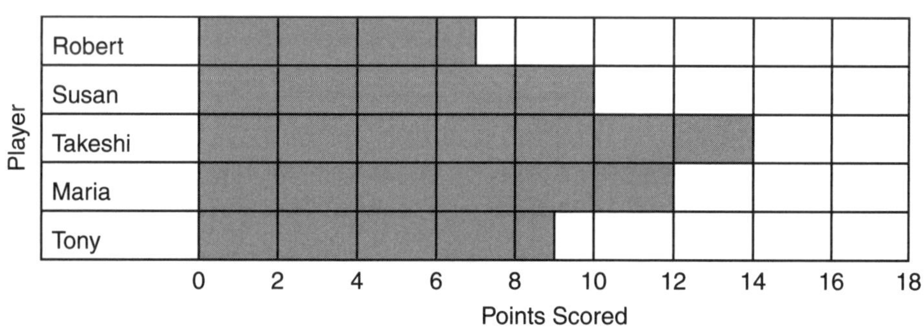

15. (Inv. 1) How many points did Maria and Susan score together?

16. (Inv. 1) Tony scored how many more points than Robert?

17. (6) What number is $\frac{1}{2}$ of \$7.50?

18. (8) What is the perimeter of this square?

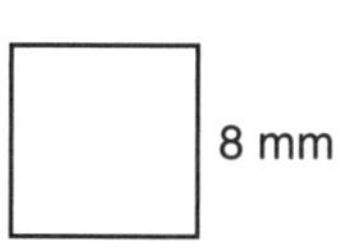

19. (10) What is the next number in this sequence?

1, 3, 9, 27, _____

20. (7) How long is this line segment?

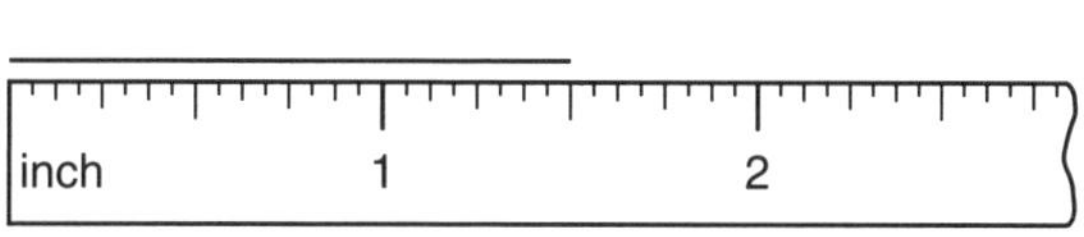

TEST

5A

Also take Facts Practice Test E (100 Multiplication Facts).

Name ____________________

1. (12) What is the difference when the sum of 20 and 10 is subtracted from the product of 20 and 10?

2. (23) In a class of 28 students there are 13 boys. What is the ratio of boys to girls in the class?

3. (12) Write the numeral for seventeen million, one hundred five thousand.

4. (16) Estimate the product of 87 and 31.

5. (21) Which of these numbers is divisible by 3?

A. 2437 B. 1632 C. 4373

6. (25) Write $\frac{12}{5}$ as a mixed number.

7. (18) What is the average of 37, 82, 63, and 58?

8. (6) What is the denominator of $\frac{11}{12}$?

9. (20) What is the greatest common factor (GCF) of 12 and 18?

10. (6) A team won 7 of its 12 games. What fraction of its games did the team win?

11. (6) What number is $\frac{1}{4}$ of 60?

12. (6) What number is $\frac{2}{3}$ of 12?

13. (24) $\frac{2}{6} + \frac{3}{6}$

14. (24) $\frac{11}{12} - \frac{4}{12}$

15. (Inv. 2) $1 - \frac{3}{5}$

16. (24) $\frac{3}{8} + \frac{5}{8}$

17. (Inv. 2) What percent of a circle is $\frac{1}{2}$ of a circle?

18. (24) $\frac{1}{8} + \frac{1}{8} + \frac{1}{8}$

19. (2) $27.50 ÷ 10

20. (Inv. 2) Compare: $\frac{1}{2} \bigcirc \frac{1}{4}$

TEST

5B

Also take Facts Practice Test E (100 Multiplication Facts).

Name ______________________

1. (12) What is the product of the sum of 20 and 10 and the difference of 20 and 10?

2. (23) In a class of 29 students there are 13 boys. What is the ratio of boys to girls in the class?

3. (12) Write the numeral for twenty-one million, two hundred eighty-six thousand.

4. (16) Round 3875 to the nearest hundred.

5. (21) Which of these numbers is divisible by 3?

A. 1367 B. 2458 C. 3624

6. (25) Write $\frac{16}{3}$ as a mixed number.

7. (18) What is the average of 3742, 4918, and 5170?

8. (6) What is the numerator of $\frac{7}{8}$?

9. (20) What is the greatest common factor (GCF) of 10 and 15?

10. (6) A team lost 4 of its 15 games. What fraction of its games did the team lose?

11. (6) What number is $\frac{1}{3}$ of 60?

12. (6) What number is $\frac{3}{4}$ of 24?

13. (24) $\frac{1}{3} + \frac{1}{3}$

14. (24) $\frac{7}{8} - \frac{4}{8}$

15. (Inv. 2) $1 - \frac{3}{8}$

16. (24) $\frac{5}{6} + \frac{1}{6}$

17. (Inv. 2) What percent of a circle is $\frac{1}{4}$ of a circle?

18. (24) $\frac{1}{12} + \frac{2}{12} + \frac{4}{12}$

19. (2) $37.50 ÷ 10

20. (Inv. 2) Compare: $\frac{1}{2} \bigcirc \frac{3}{4}$

TEST

6A

Also take Facts Practice Test E (100 Multiplication Facts).

Name ______________________________

1. *(25)* A 27-inch length of ribbon was cut into 4 equal lengths. How long was each piece of ribbon?

2. *(29)* In a class of 30 students there are 16 girls. What is the ratio of boys to girls in the class?

3. *(16)* Estimate the product of 68 and 21.

4. *(2)* $\frac{567}{9}$

5. *(18)* What is the average of 68, 72, and 82?

6. *(19)* List the six whole-number factors of 20.

7. *(20)* What is the GCF of 24 and 16?

8. *(27)* Which of these words names the distance across a circle through its center?

A. radius B. diameter C. circumference

9. *(22)* If $\frac{3}{4}$ of the 12 doughnuts were eaten, how many were eaten?

10. *(5, 24)* $\left(\frac{2}{7} + \frac{2}{7}\right) - \frac{4}{7}$

11. *(Inv. 2)* $1 - \frac{1}{3}$

12. *(29)* $\frac{1}{2} \times \frac{3}{4}$

13. *(25, 29)* Convert $\frac{10}{6}$ to a mixed number and reduce the fraction.

14. *(26)* Add and convert: $3\frac{2}{3} + 2\frac{2}{3}$

15. *(24, 29)* Add and reduce: $\frac{3}{8} + \frac{3}{8}$

16. *(30)* What is the least common multiple (LCM) of 4 and 6?

17. *(8)* The perimeter of a square is 36 in. How long is each side?

18. *(30)* $\frac{2}{3} \times \square = 1$

19. *(10)* What are the next three numbers in this sequence?

8, 16, 24, 32, _____, _____, _____

20. *(28)* Name the obtuse angle in triangle *ABC*.

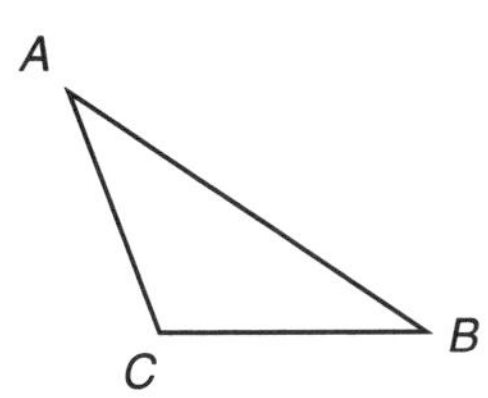

TEST

6B

Also take Facts Practice Test E (100 Multiplication Facts).

Name ______________________________

1. (25) A 29-inch length of ribbon was cut into 4 equal lengths. How long was each piece of ribbon?

2. (29) In a class of 30 students there are 18 girls. What is the ratio of boys to girls in the class?

3. (16) Estimate the difference of 6084 and 3973 to the nearest thousand.

4. (2) $\frac{432}{18}$

5. (18) What is the average of 143, 98, and 128?

6. (19) List the four whole-number factors of 33.

7. (20) What is the greatest common factor of 27 and 45?

8. (27) Which of these words names the distance around a circle?

A. radius B. diameter C. circumference

9. (22) A score is 20. How many is $\frac{3}{4}$ of a score?

10. (5, 24) $\left(\frac{3}{5} - \frac{2}{5}\right) + \frac{1}{5}$

11. (Inv. 2) $1 - \frac{1}{4}$

12. (29) $\frac{1}{3} \times \frac{2}{5}$

13. (25, 29) Convert $\frac{12}{8}$ to a mixed number and reduce the fraction.

14. (26) Add and convert: $4\frac{4}{5} + 3\frac{3}{5}$

15. (24, 29) Subtract and reduce: $\frac{5}{6} - \frac{1}{6}$

16. (30) What is the least common multiple (LCM) of 6 and 8?

17. (8) The perimeter of a square is 60 cm. How long is each side?

18. (30) $\frac{3}{4} \times \square = 1$

19. (10) What is the seventh number in this sequence?

7, 14, 21, 28, …

20. (28) Name the obtuse angle in triangle *ABC*.

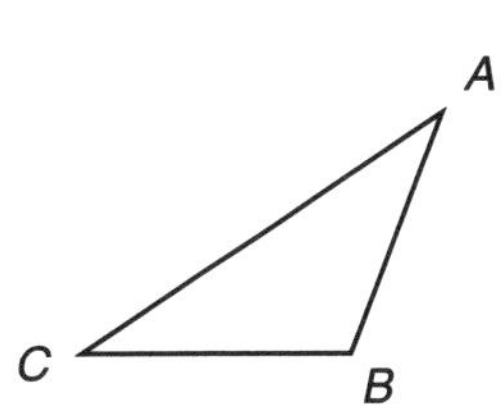

TEST

7A

Also take Facts Practice Test F (90 Division Facts).

Name ______________________________

1. One mile is 1760 yards. Round 1760 to the nearest hundred.
(16)

2. What is the reciprocal of $\frac{2}{3}$?
(30)

3. What is the perimeter of a square that has sides 8 inches long?
(8)

4. How many square tiles with sides 1 ft long would be needed to cover the floor of a 10-ft-by-12-ft room?
(31)

5. Which digit in 5.432 is in the hundredths place?
(34)

6. Write the decimal numeral for twelve thousandths.
(35)

7. Write $(5 \times 100) + (6 \times 1)$ in standard notation.
(32)

8. The diameter of the bicycle tire is 24 inches. What is the radius of the tire?
(27)

9. Write 0.47 as a common fraction.
(35)

10. Compare: $\frac{3}{4} \bigcirc \frac{4}{3}$
(Inv. 2)

11. Write $\frac{35}{100}$ in decimal form.
(35)

12. Add and convert to a mixed number: $\frac{3}{5} + \frac{4}{5}$
(24, 25)

13. $\frac{2}{3} \times \frac{3}{5}$
(29)

14. $18 \times 20 \div 12$
(5)

15. What number is $\frac{3}{4}$ of 36?
(22)

16. Subtract and reduce: $\frac{5}{8} - \frac{1}{8}$
(24, 29)

Find each missing number:

17. $w - 10 = 100$
(3)

18. $3 + N = 4\frac{1}{3}$
(3, 26)

19. Write 80% as a reduced fraction.
(33)

20. Four fifths of the 40 answers were correct. How many answers were correct?
(22)

TEST 7B

Also take Facts Practice Test F (90 Division Facts).

Name ______________________________

1. One mile is 5280 feet. Round 5280 to the nearest hundred.
(16)

2. What is the reciprocal of $\frac{5}{6}$?
(30)

3. What is the perimeter of a square that has sides 10 centimeters long?
(8)

4. How many square tiles with sides 1 ft long would be needed to cover the floor of a 10-ft-by-15-ft room?
(31)

5. Which digit in 57.894 is in the tenths place?
(34)

6. Write the decimal numeral for one and five tenths.
(35)

7. Write $(5 \times 1000) + (7 \times 10)$ in standard notation.
(32)

8. The diameter of the bicycle tire is 28 inches. What is the radius of the tire?
(27)

9. Write 0.7 as a common fraction.
(35)

10. Compare: $\frac{6}{5} \bigcirc \frac{5}{6}$
(Inv. 2)

11. Write $\frac{3}{100}$ in decimal form.
(35)

12. Add and convert to a mixed number: $\frac{3}{7} + \frac{6}{7}$
(24, 25)

13. $\frac{4}{5} \times \frac{3}{4}$
(29)

14. $23 \times (500 \div 20)$
(5)

15. What number is $\frac{5}{6}$ of 36?
(22)

16. Subtract and reduce: $\frac{5}{6} - \frac{1}{6}$
(24, 29)

Find each missing number:

17. $w - 50 = 500$
(3)

18. $3 + N = 5\frac{2}{3}$
(3, 26)

19. Write 60% as a reduced fraction.
(33)

20. Three fourths of the 40 answers were correct. How many answers were correct?
(22)

TEST 8A

Also take Facts Practice Test G (30 Fractions to Reduce).

Name ____________________

1. *(Inv. 3)* Which of these angles could measure 60°?

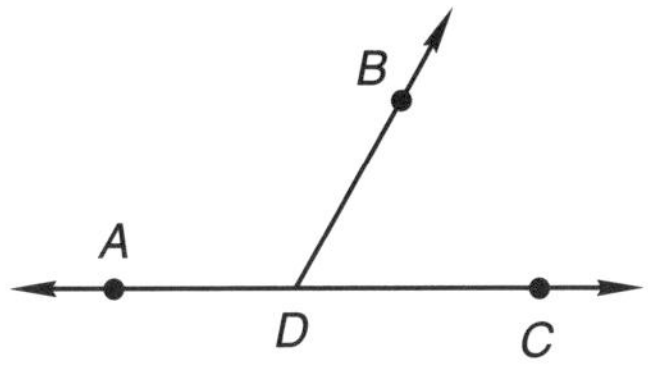

2. *(12)* Which digit in 987,654,321,000 is in the ten-millions place?

3. *(29)* Maria correctly answered 16 of the 20 questions. What fraction of the questions did she answer correctly?

4. *(16)* Estimate the sum of 6107 and 2960 to the nearest thousand.

5. *(38)* $4^2 + \sqrt{9}$

6. *(18)* Dan's three test scores were 78, 75, and 90. What was the average of his three scores?

7. *(20)* What is the largest number that is a factor of both 12 and 20?

8. *(33)* Sixty percent of the answers were right. What fraction of the answers were right?

9. *(26)* $1\frac{2}{3} + 2\frac{2}{3}$

10. *(Inv. 2)* $1 - \frac{5}{8}$

11. *(29)* What is the product of $\frac{1}{3}$ and $\frac{3}{4}$?

12. *(32)* The movie starts at 3:15 p.m. and ends at 5:08 p.m. How long is the movie?

13. *(30)* What is the least common multiple (LCM) of 6 and 8?

14. *(31)* One side of a square is 10 inches long. How many one-square-inch tiles are needed to cover its area?

15. *(34)* Which digit in 12.34 has the same place value as the 6 in 5.678?

16. *(35)* Write the decimal numeral for ten and five hundredths.

17. *(37)* $0.7 + 0.8$

18. *(38)* $0.2 + 0.34 + 5$

19. *(37)* $0.15 - 0.06$

20. *(39)* 0.15×0.3

TEST

8B

Also take Facts Practice Test G (30 Fractions to Reduce).

Name ____________________

1. Which of these angles could measure 120°?
(Inv. 3)

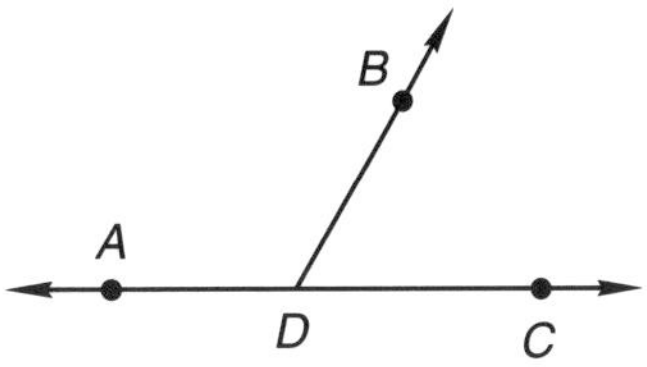

2. Which digit in 437,286,190 is in the hundred-thousands place?
(12)

3. Marco correctly answered 15 of the 20 questions. What fraction of the questions did he answer correctly?
(29)

4. Estimate the sum of 8972 and 4106 to the nearest thousand.
(16)

5. $5^2 + \sqrt{16}$
(38)

6. Juanita's three test scores were 81, 84, and 84. What was the average of her three scores?
(18)

7. What is the greatest common factor (GCF) of 21 and 35?
(20)

8. Forty percent of the students were boys. What fraction of the students were boys?
(33)

9. $3\frac{3}{4} + 1\frac{1}{4}$
(26)

10. $5 - 1\frac{1}{3}$
(Inv. 2)

11. What is the product of $\frac{1}{2}$ and $\frac{2}{3}$?
(29)

12. School starts at 8:30 a.m. and ends at 2:45 p.m. How many hours and minutes are there from the start to the end of school?
(32)

13. What is the least common multiple (LCM) of 3 and 6?
(30)

14. A rectangle is 8 inches long and 6 inches wide. How many one-square-inch tiles are needed to cover its area?
(31)

15. Which digit in 12.345 is in the hundredths place?
(34)

16. Write the decimal numeral for three and twelve hundredths.
(35)

17. $0.3 + 0.4 + 0.5$
(37)

18. $6 + 4.2 + 0.36$
(38)

19. $0.23 - 0.19$
(37)

20. 0.4×0.12
(39)

TEST

9A

Also take Facts Practice Test G (30 Fractions to Reduce).

Name ______________________________

1. (12) What is the difference when the sum of 7 and 8 is subtracted from the product of 10 and 6?

2. (30) How many $\frac{3}{4}$'s are in 1?

3. (8) If the perimeter of a square is 36 inches, how long is each side?

4. (31) What is the area of this rectangle?

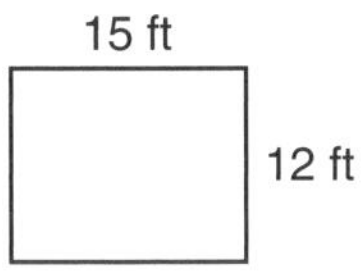

5. (35) Write the decimal numeral for ten and twelve hundredths.

6. (41) An \$80 pair of shoes is on sale for 25% off the regular price. How much money is 25% of \$80?

7. (22) How much is $\frac{3}{4}$ of two dozen?

8. (Inv. 2) Which percent best describes the shaded portion of this circle?

A. 30% B. 50%
C. 70% D. 90%

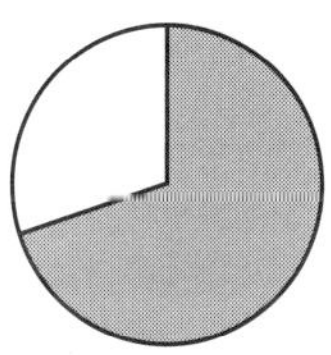

9. (34) Which digit in 123.456 is in the tenths place?

10. (41) Write 7% as a fraction and as a decimal numeral.

11. (20) What is the greatest common factor (GCF) of 12 and 8?

12. (38) $2\frac{1}{3} + \sqrt{9}$

13. (29) $\frac{1}{2} \times \frac{2}{3}$

14. (38) $0.43 + 6.7 + 8$

15. (40, 43) Solve: $3.27 + n = 5$

16. (39) 0.34×0.26

17. (26) $4\frac{2}{3} - 3\frac{1}{3}$

18. (42) $\frac{2}{3} = \frac{?}{12}$

19. (45) $4\overline{)0.7}$

20. (27) Which segment in the circle at right is a diameter?

A. $\overline{SQ}$ B. $\overline{MS}$
C. $\overline{SR}$ D. $\overline{MR}$

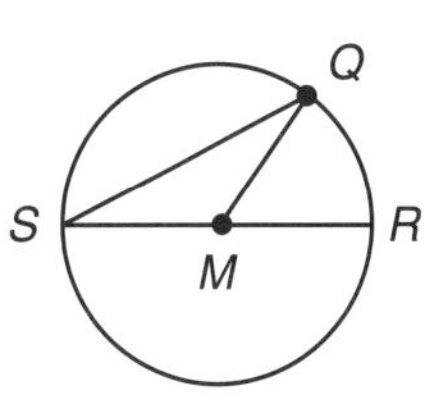

TEST 9B

Also take Facts Practice Test G (30 Fractions to Reduce).

Name ____________________

1. (12) What is the product of the sum of 8 and 2 and the difference of 8 and 2?

2. (30) How many $\frac{2}{3}$'s are in 1?

3. (8) If the perimeter of a square is 100 centimeters, how long is each side?

4. (31) What is the area of this rectangle?

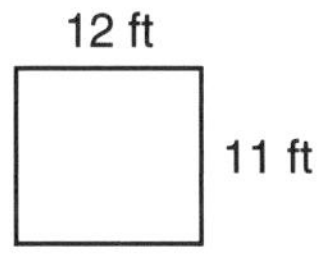

5. (35) Write the decimal numeral for twelve and five hundredths.

6. (41) A \$60 pair of shoes is on sale for 25% off the regular price. How much money is 25% of \$60?

7. (22) How much is $\frac{2}{3}$ of two dozen?

8. (Inv. 2) Which percent best describes the shaded portion of this circle?

A. 30% B. 50%
C. 70% D. 90%

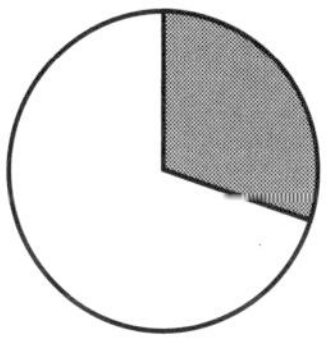

9. (34) Which digit in 14.376 has the same place value as the 5 in 2.05?

10. (41) Write 9% as a fraction and as a decimal numeral.

11. (30) What is the least common multiple (LCM) of 12 and 8?

12. (38) $1\frac{3}{5} + \sqrt{4}$

13. (29) $\frac{1}{3} \times \frac{3}{4}$

14. (38) $6.2 + 7 + 0.48$

15. (40, 43) Solve: $0.34 + m = 2$

16. (39) 0.23×0.12

17. (26) $3\frac{3}{5} + 3\frac{2}{5}$

18. (42) $\frac{2}{3} = \frac{?}{9}$

19. (45) $8\overline{)3.0}$

20. (27) Which segment in the circle at right is not a radius?

A. $\overline{MS}$ B. $\overline{MR}$
C. $\overline{MQ}$ D. $\overline{SQ}$

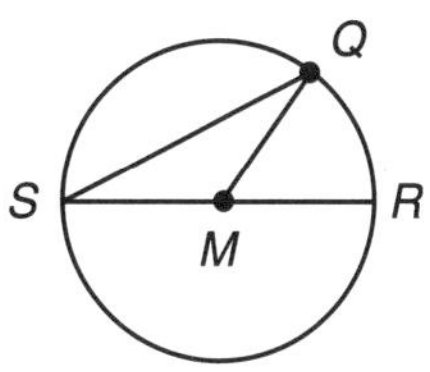

TEST 10A

Also take Facts Practice Test H (72 Multiplication and Division Facts).

Name ____________________

1. (12) What is the difference when the product of 0.3 and 0.4 is subtracted from the sum of 0.3 and 0.4?

2. (12) Which digit in 12,345,678,900 is in the hundred-millions place?

3. (41) The sales-tax rate is 8%. How much money is 8% of $16?

4. (16) Estimate the sum of 5867 and 2106 to the nearest thousand.

5. (2) $\dfrac{40{,}100}{20}$

6. (18) What is the average of 96, 100, and 92?

7. (24, 29) Add and reduce: $\dfrac{7}{24} + \dfrac{11}{24}$

8. (22) How much is $\dfrac{5}{6}$ of two dozen?

9. (42) Write $\dfrac{1}{4}$ as a fraction with 8 as the denominator. Then add the fraction to $\dfrac{3}{8}$. What is the sum?

10. (30) What is the least common multiple (LCM) of 2, 3, and 4?

11. (31) What is the area of this square?

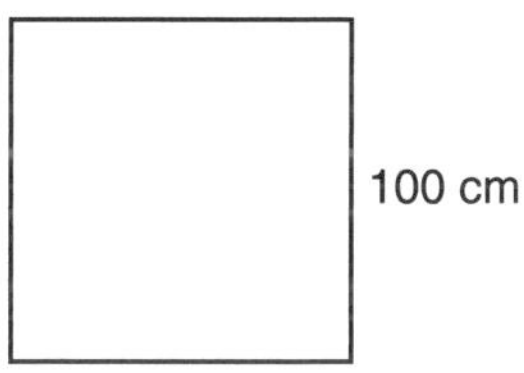

12. (47) The diameter of a pizza is 10 inches. What is the circumference of the pizza? (Use 3.14 for π.)

13. (35) Write the decimal numeral for ten and two tenths.

14. (40, 43) Solve: $1 - x = 0.23$

15. (38) $5.62 + 4 + 0.5$

16. (39) $0.2 \times 0.3 \times 0.4$

17. (45) $0.123 \div 3$

18. (49) $0.45 \div 0.3$

19. (26) $6\frac{1}{5} - 1\frac{3}{5}$

20. (46) Mentally calculate this product: 12.5×100

TEST

10B

Also take Facts Practice Test H (72 Multiplication and Division Facts).

Name ____________________

1. (12) What is the difference when the product of $\frac{1}{3}$ and $\frac{2}{3}$ is subtracted from the sum of $\frac{1}{3}$ and $\frac{2}{3}$?

2. (12) Which digit in 987,654,321,000 is in the billions place?

3. (41) The sales-tax rate is 8%. How much money is 8% of $14?

4. (16) Estimate the product of 48 and 21.

5. (2) $\frac{2508}{12}$

6. (18) What is the average of 47, 38, 56, and 43?

7. (24, 29) Add and reduce: $\frac{11}{36} + \frac{13}{36}$

8. (22) How much is $\frac{3}{4}$ of four dozen?

9. (12) Write $\frac{1}{4}$ as a fraction with a denominator of 8. Then add the fraction to $\frac{5}{8}$. What is the sum?

10. (30) What is the least common multiple (LCM) of 2, 4, and 8?

11. (31) What is the area of this rectangle?

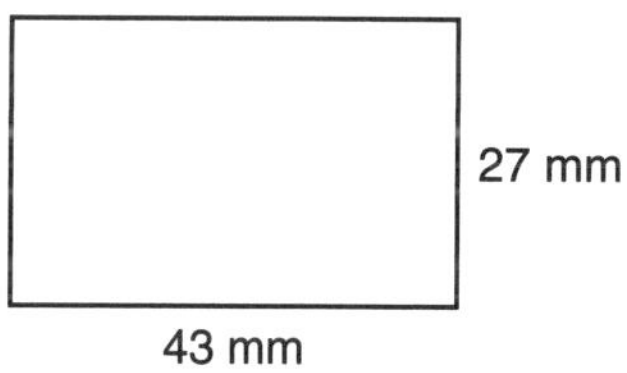

12. (47) The diameter of a circular wading pool is 10 ft. What is the circumference of the pool? (Use 3.14 for π.)

13. (35) Write the decimal numeral for twenty-one and five hundredths.

14. (40, 43) Solve: $1 - x = 0.01$

15. (38) $4.6 + 0.47 + 12$

16. (39) 0.12×0.13

17. (45) $0.456 \div 6$

18. (49) $0.123 \div 0.03$

19. (26) $5\frac{1}{3} - 2\frac{2}{3}$

20. (46) Mentally calculate this product: 3.5×100

TEST

11A

Also take Facts Practice Test I (28 Improper Fractions to Simplify).

Name ______________________

1. (47) Raul used a compass to draw a circle with a radius of 4 cm. What was the circumference of the circle? (Use 3.14 for π.)

2. (39) What is the product of 1.5 and 0.025?

Refer to the rectangle on the right for problems 3–5.

3. (8) What is the perimeter of the rectangle?

4. (31) What is the area of the rectangle?

5. (23) What is the ratio of the length of the rectangle to its width?

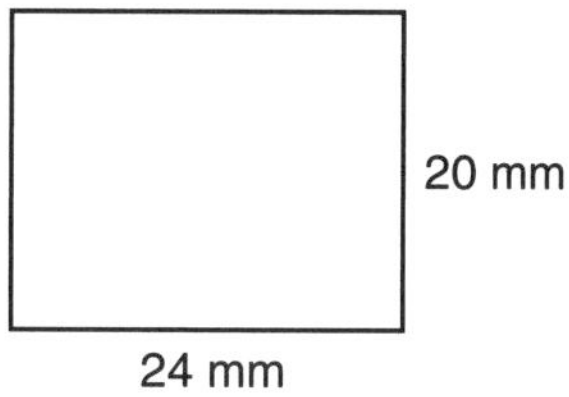

6. (41) Hannah answered 30% of the 50 questions correctly. How many questions did she answer correctly?

7. (43) Solve: $7 - n = 4.7$

8. (44) Compare: 0.60 ◯ 0.6

9. (35) Write the decimal numeral for eighty and eight hundredths.

10. (51) Round 2.8735 to the nearest tenth.

11. (26) $3\frac{2}{3} + 1\frac{2}{3}$

12. (48) $6\frac{1}{3} - 4\frac{2}{3}$

13. (29) $\frac{3}{8} \cdot \frac{4}{9}$

14. (43) Solve: $x + \frac{2}{3} = 1\frac{1}{3}$

15. (52) $3.75 \div 10^2$

16. (49) $0.4\overline{)0.015}$

17. (38, 49) $\sqrt{36} \div 0.6$

18. (55) $\frac{3}{4} + \frac{1}{8}$

19. (54) Reduce: $\dfrac{2 \cdot 2 \cdot 3 \cdot 3 \cdot 5}{2 \cdot 3 \cdot 5 \cdot 7}$

20. (50) To which decimal number is the arrow pointing?

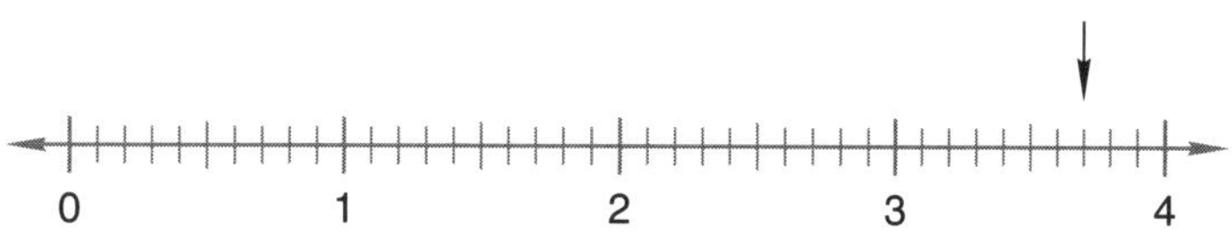

TEST 11B

Also take Facts Practice Test I (28 Improper Fractions to Simplify).

Name ______________________

1. (47) Rita used a compass to draw a circle with a radius of 3 cm. What was the circumference of the circle? (Use 3.14 for π.)

2. (39) What is the product of 0.81 and 0.012?

Refer to the rectangle on the right for problems 3–5.

3. (8) What is the perimeter of the rectangle?

4. (31) What is the area of the rectangle?

5. (23) What is the ratio of the length of the rectangle to its width?

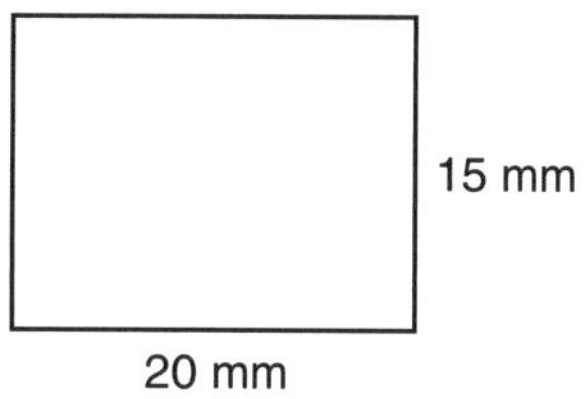

6. (41) Curly answered 20% of the 50 questions correctly. How many questions did he answer correctly?

7. (43) Solve: $6 - w = 3.1$

8. (44) Compare: 1.23 ◯ 0.140

9. (35) Write the decimal numeral for twenty and two tenths.

10. (51) Round 6.7347 to the nearest hundredth.

11. (26) $3\frac{3}{4} + 2\frac{1}{4}$

12. (48) $5\frac{1}{5} - 1\frac{3}{5}$

13. (29) $\frac{3}{4} \cdot \frac{2}{5}$

14. (43) Solve: $x + \frac{3}{5} = 1\frac{2}{5}$

15. (52) $0.45 \div 10^2$

16. (49) $0.2\overline{)0.015}$

17. (38, 49) $\sqrt{4} \div 0.4$

18. (55) $\frac{1}{4} + \frac{3}{8}$

19. (54) Reduce: $\dfrac{2 \cdot 3 \cdot 3 \cdot 5}{2 \cdot 2 \cdot 3 \cdot 5 \cdot 5}$

20. (50) To which decimal number is the arrow pointing?

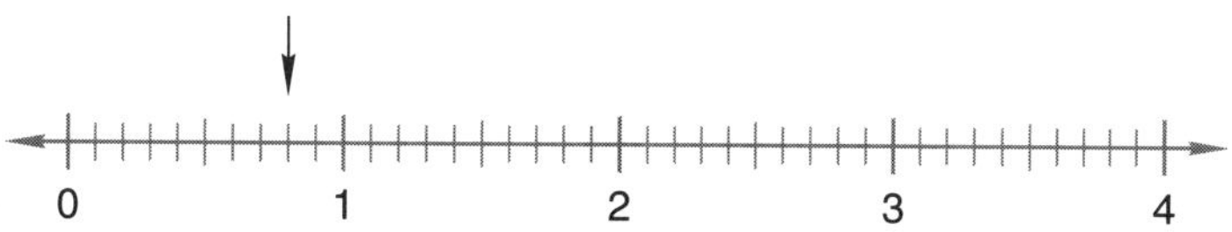

TEST

12A

Also take Facts Practice Test H (72 Multiplication and Division Facts).

Name ____________________

1. (58) If a number cube is rolled, what is the probability that the number rolled will be greater than 4?

2. (60) What is the perimeter of a regular hexagon if each side is 8 inches long?

3. (31) How many square centimeters are needed to cover this rectangle?

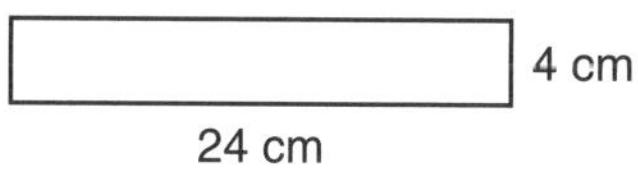

4. (34) Which digit in 42.36 has the same place value as the 7 in 158.7?

5. (38) $3 + 4.5 + 0.23$

6. (40) $1 - 0.29$

7. (39) 0.3×0.27

8. (47) The diameter of an automobile's tire is 30 inches. How far down the road will the tire travel if it makes one full turn? (Use 3.14 for π and round your answer to the nearest inch.)

9. (19) Which of these numbers is not a prime number?

A. 3 B. 5 C. 7 D. 9

10. (23) The team won 10 of its 30 games and lost the rest. What was the team's win-loss ratio?

11. (36) $5 - 2\frac{1}{3}$

12. (45) $0.45 \div 9$

13. (49) $1.2 \div 0.04$

14. (46) 12.34×10

15. (56) $\frac{1}{2} + \frac{3}{5}$

16. (56) $\frac{4}{5} - \frac{1}{4}$

17. (56) Compare: $\frac{3}{4} \bigcirc \frac{2}{3}$

18. (59) $3\frac{1}{2} + 2\frac{2}{3}$

19. (48) $7\frac{1}{4} - 1\frac{3}{4}$

20. (41) The sales-tax rate was 8%. Benito bought a model train for \$8.29. How much was the tax on the train?

TEST

12B

Also take Facts Practice Test H (72 Multiplication and Division Facts).

Name ______________________

1. (58) If a number cube is rolled, what is the probability that the number rolled will be less than 4?

2. (60) What is the perimeter of a regular hexagon if each side is 7 inches long?

3. (31) How many square centimeters are needed to cover this rectangle?

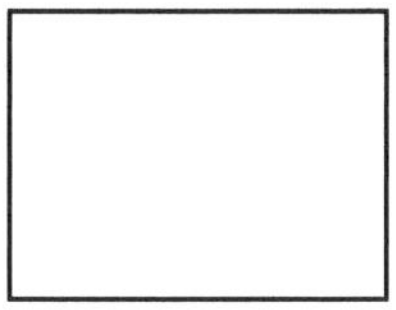

4. (34) Which digit in 5.6789 is in the thousandths place?

5. (38) 6.2 + 0.56 + 12

6. (40) 2.5 − 0.25

7. (39) 1.4×0.06

8. (47) The diameter of a truck's tire is 50 inches. How far down the road will the tire travel if it makes one full turn? (Use 3.14 for π and round your answer to the nearest inch.)

9. (19) Which of these numbers is not a prime number?

A. 2 B. 3 C. 4 D. 5

10. (23) The team won 12 of its 30 games and lost the rest. What was the team's win-loss ratio?

11. (36) $1 - \frac{5}{8}$

12. (45) $1.2 \div 5$

13. (49) $24 \div 0.6$

14. (52) $12.34 \div 10$

15. (56) $\frac{3}{4} + \frac{1}{3}$

16. (56) $\frac{2}{3} - \frac{1}{5}$

17. (56) Compare: $\frac{5}{8} \bigcirc \frac{3}{4}$

18. (59) $4\frac{3}{4} + 2\frac{1}{2}$

19. (48w) $8\frac{1}{3} - 2\frac{2}{3}$

20. (41) The sales-tax rate was 8%. Bethany bought a book for \$9.49. How much was the tax on the book?

TEST

13A

Also take Facts Practice Test J (24 Mixed Numbers to Write as Improper Fractions).

Name ______________________

1. (31) The perimeter of a square is 8 feet. What is its area?

2. (42) By what fraction equal to 1 should $\frac{2}{3}$ be multiplied to form $\frac{8}{12}$?

3. (58) The face of this spinner is divided into six congruent sectors. If the spinner is spun once, what is the probability that it will stop on an even number?

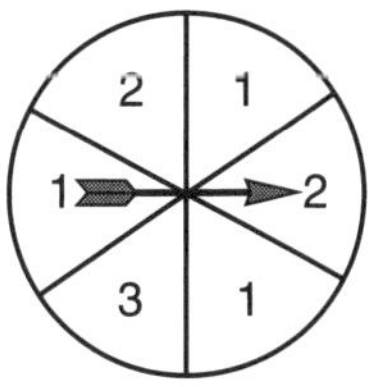

4. (27) The diameter of a big circle is 42 feet. What is the ratio of the circle's radius to its diameter?

5. (64) Which quadrilateral is not a parallelogram?

A.  B.  C. 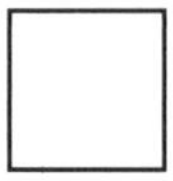 D. 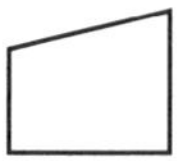

6. (65) Forty is a composite number. Use division by primes to find the prime factorization of 40.

7. (56) Compare: $\frac{3}{5} \bigcirc \frac{2}{3}$

8. (41) An item costs $12.95 and the sales-tax rate is 8%. What is the total price of the item including sales tax?

9. (1) $6 − 6¢

10. (38) $3.7 + 12 + 4.31$

11. (40) $7.4 - 4.74$

12. (39) 0.18×0.27

13. (45) $5\overline{)0.14}$

14. (49) $0.6\overline{)9}$

15. (63) $3\frac{5}{6} - 1\frac{1}{2}$

16. (59) $1\frac{2}{3} + 1\frac{5}{6}$

17. (61) $\frac{1}{2} + \frac{3}{4} + \frac{3}{8}$

18. (43) Solve: $6\frac{1}{3} - m = 3$

19. (Inv. 6) A cube has how many vertices?

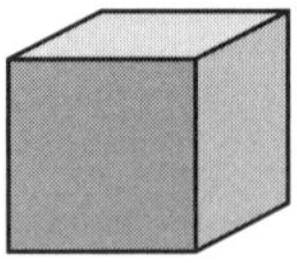

20. (62) Write $3\frac{3}{4}$ as an improper fraction. Then multiply the improper fraction by $\frac{1}{6}$. Write the product as a reduced fraction.

TEST 13B

Also take Facts Practice Test J (24 Mixed Numbers to Write as Improper Fractions).

Name ______________________

1. (31) If the perimeter of a square is 16 inches, what is its area?

2. (42) By what fraction equal to 1 should $\frac{3}{4}$ be multiplied to form $\frac{9}{12}$?

3. (58) The face of this spinner is divided into six congruent sectors. If the spinner is spun once, what is the probability that it will stop on an odd number?

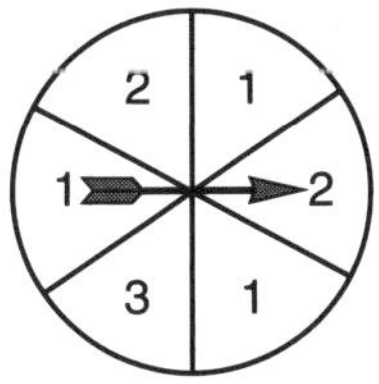

4. (27) The diameter of a big circle is 36 feet. What is the ratio of the circle's radius to its diameter?

5. (64) Which quadrilateral is not a parallelogram?

A. 

B. 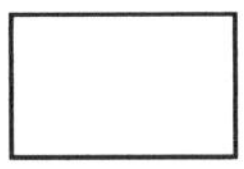

C. 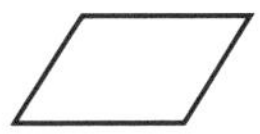

D. 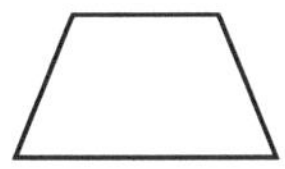

6. (65) Fifty is a composite number. Use a factor tree to find the prime factorization of 50.

7. (56) Compare: $\frac{5}{6} \bigcirc \frac{5}{8}$

8. (41) An item costs \$14.85 and the sales-tax rate is 7%. What is the total price of the item including sales tax?

9. (1) \$10 − 10¢

10. (38) $4.87 + 16 + 12.8$

11. (40) $4.7 - 1.79$

12. (39) $0.5 \times 0.6 \times 0.7$

13. (45) $4\overline{)0.15}$

14. (49) $0.6\overline{)15}$

15. (63) $4\frac{2}{3} - 1\frac{1}{6}$

16. (59) $3\frac{1}{2} + 4\frac{5}{6}$

17. (61) $\frac{1}{2} + \frac{1}{3} + \frac{1}{6}$

18. (43) Solve: $5\frac{1}{4} - n = 2$

19. (Inv. 6) A cube has how many edges?

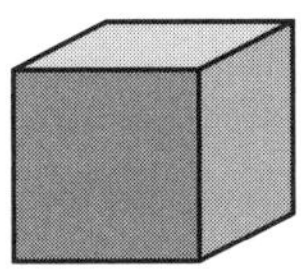

20. (62) Write $2\frac{2}{3}$ as an improper fraction. Then multiply the improper fraction by $\frac{1}{4}$. Write the product as a reduced fraction.

TEST

14A

Also take Facts Practice Test I (28 Improper Fractions to Simplify).

Name ______________________________

1. (34) Which digit is in the hundredths place in 4567.891?

2. (31) The perimeter of a square is 44 centimeters. What is its area?

3. (40) $10 - (0.25 + 1)$

4. (39) 0.12×0.12

5. (42) $\frac{11}{20} = \frac{?}{100}$

6. (38, 45) $0.45 \div \sqrt{36}$

7. (64) Which figure is not a quadrilateral?

A. 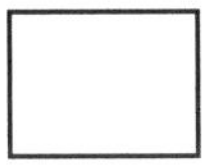

B. 

C. 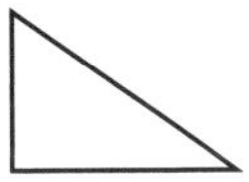

D. 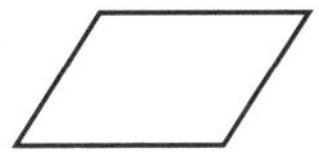

8. (44) Compare: $0.2 + 0.2 \bigcirc 0.2 \times 0.2$

9. (41) What is the total price of a $24.95 item plus 6% sales tax?

10. (46) 4.5×100

11. (49) $1.2 \div 0.08$

12. (63) $4\frac{1}{5} - 3\frac{1}{2}$

13. (61) $1\frac{1}{2} + 2\frac{1}{4} + 3\frac{3}{8}$

14. (Inv. 6) How many small cubes are used to form this larger cube?

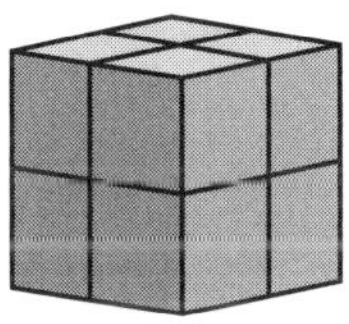

15. (60) Each side of a regular hexagon is 12 mm. What is the perimeter of the hexagon?

16. (67) Write the prime factorization of the numerator and denominator of $\frac{24}{30}$. Then reduce.

17. (70) $2\frac{1}{4} \times 1\frac{2}{3}$

18. (68) $1\frac{3}{4} \div 2$

19. (7) Which figure is a line segment?

A.

B.

C.

20. (58) A bag contains 3 red marbles, 4 white marbles, and 5 blue marbles. If one marble is drawn from the bag, what is the probability that the marble will be red?

TEST

14B

Also take Facts Practice Test I (28 Improper Fractions to Simplify).

Name ______________________

1. Which digit is in the hundredths place in 9876.543?
(34)

2. The perimeter of a square is 48 millimeters. What is its area?
(31)

3. $2 - (0.6 + 0.7)$
(40)

4. 0.12×0.13
(39)

5. $\frac{9}{20} = \frac{?}{100}$
(42)

6. $0.28 \div \sqrt{64}$
(38, 45)

7. Which figure is a quadrilateral?
(64)

A. 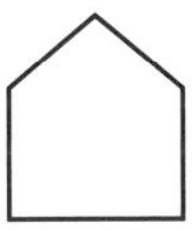

B. 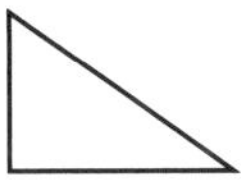

C. 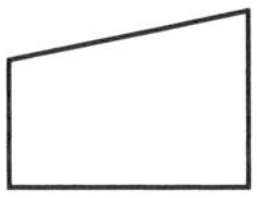

D. 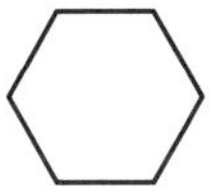

8. Compare: $0.2 + 0.2 \bigcirc 0.2 \div 0.2$
(44)

9. What is the total price of an $18.97 item plus 8% sales tax?
(41)

10. $4.5 \div 100$
(52)

11. $3 \div 0.4$
(49)

12. $5\frac{1}{8} - 3\frac{1}{2}$
(63)

13. $6\frac{1}{3} \mid 2\frac{1}{2} \mid 5\frac{1}{6}$
(61)

14. How many small cubes are used to form this larger cube?
(Inv. 6)

15. The perimeter of a regular octagon is 48 cm. How long is each side?
(60)

16. Write the prime factorization of the numerator and denominator of $\frac{24}{60}$. Then reduce.
(67)

17. $1\frac{1}{3} \times 1\frac{3}{4}$
(70)

18. $1\frac{1}{2} \div 2\frac{1}{2}$
(68)

19. Which figure is a ray?
(7)

A.

B.

C.

20. A bag contains 3 red marbles, 4 white marbles, and 5 blue marbles. If one marble is drawn from the bag, what is the probability that the marble will be white?
(58)

TEST

15A

Also take Facts Practice Test J (24 Mixed Numbers to Write as Improper Fractions).

Name ______________________________

1. (71) What is the area of this parallelogram?

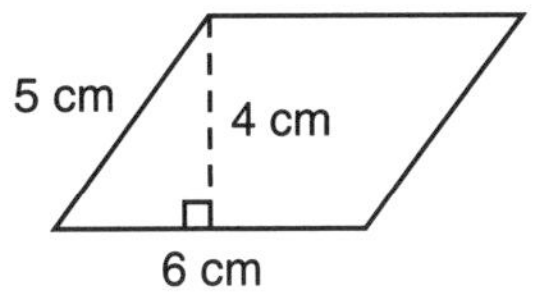

2. (41) Quan ordered a \$6.80 meal. The tax rate was 7%. He paid with a \$10 bill. How much money should he get back?

3. (56) Compare: $\frac{3}{4} \bigcirc \frac{11}{16}$

Refer to the coordinate plane on the right for problems 4 and 5.

4. (Inv. 7) Which point has the coordinates (2, –3)?

5. (Inv. 7) What are the coordinates of point *C?*

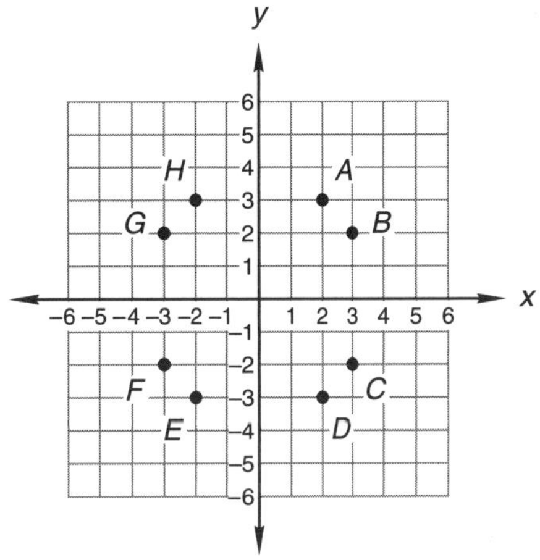

6. (60) What is the name of a five-sided polygon?

A. hexagon B. pentagon C. pyramid

7. (67) Write the prime factorizations of 45 and 54 to reduce $\frac{45}{54}$.

8. (74) Write $\frac{1}{4}$ in decimal form.

9. (41) Write 12% as a decimal numeral and as a reduced fraction.

10. (30) What is the least common multiple (LCM) of 6, 8, and 12?

11. (40) $6 - (3.7 + 0.48)$

12. (39) 1.2×0.3

13. (49) $5 \div 0.4$

14. (52) $14.5 \div 100$

15. (61) $\frac{1}{2} + \frac{2}{3} + \frac{3}{4}$

16. (59) $3\frac{3}{5} + 6\frac{1}{2}$

17. (63) $6\frac{1}{4} - 3\frac{7}{8}$

18. (70) $3\frac{1}{3} \times 1\frac{1}{5}$

19. (68) $5 \div 1\frac{2}{3}$

20. (43) Solve: $12.6 - w = 8.3$

TEST

15B

Also take Facts Practice Test J (24 Mixed Numbers to Write as Improper Fractions).

Name ______________________

1. (71) What is the area of this parallelogram?

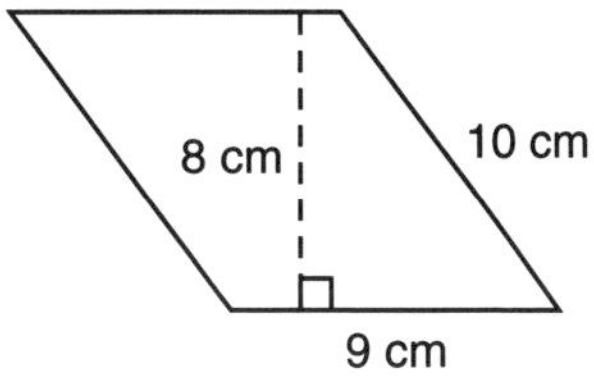

2. (41) Kim ordered a $5.90 meal. The tax rate was 8%. She paid with a $10 bill. How much money should she get back?

3. (56) Compare: $\frac{3}{5} \bigcirc \frac{13}{20}$

Refer to the coordinate plane on the right for problems 4 and 5.

4. (Inv. 7) Which point has the coordinates (–2, 3)?

5. (Inv. 7) What are the coordinates of point *D?*

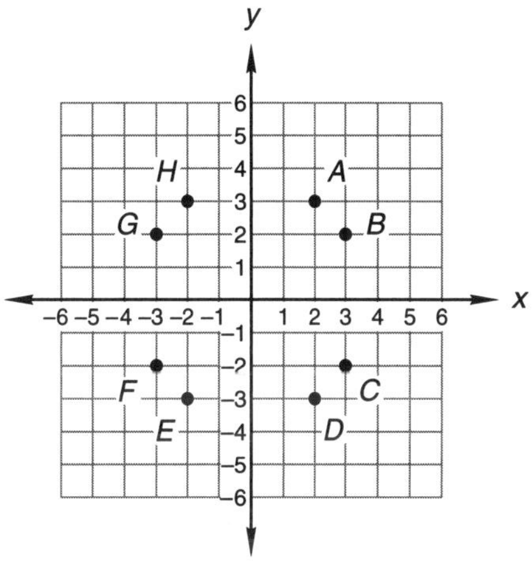

6. (60) An octagon is a polygon with how many sides?

7. (67) Write the prime factorizations of 54 and 81 to reduce $\frac{54}{81}$.

8. (74) Write $\frac{4}{5}$ in decimal form.

9. (41) Write 15% as a decimal numeral and as a reduced fraction.

10. (30) What is the least common multiple (LCM) of 4, 8, and 12?

11. (40) $8 - (4.3 + 0.75)$

12. (39) 0.45×0.2

13. (49) $4 \div 0.5$

14. (52) $12.5 \div 100$

15. (61) $\frac{1}{2} + \frac{3}{4} + \frac{3}{8}$

16. (59) $7\frac{1}{2} + 3\frac{5}{6}$

17. (63) $5\frac{2}{3} - 1\frac{3}{4}$

18. (70) $4\frac{1}{2} \times 2\frac{2}{3}$

19. (68) $4 \div 1\frac{1}{3}$

20. (43) Solve: $14.5 - m = 3.7$

TEST

16A

Also take Facts Practice Test K (Linear Measurement).

Name ______________________

1. A quart is what fraction of a gallon?
(78)

2. Sue correctly answered 42 of the 50 questions. What percent of the questions did she answer correctly?
(75)

3. Figure *ABCD* is a rectangle. Segment *AB* is 6 cm long, segment *BC* is 8 cm long, and segment *AC* is 10 cm long. What is the area of triangle *ABC?*
(79)

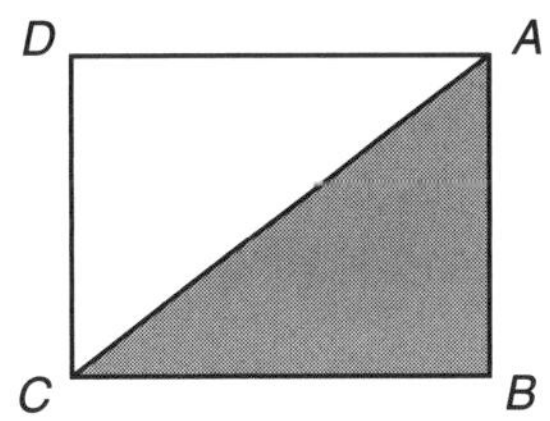

The coordinates of the four vertices of a rectangle are (1, –1), (4, –1), (4, 1), and (1, 1). Use this information for problems 4 and 5.

4. The perimeter of the rectangle is how many units?
(Inv. 7)

5. The area of the rectangle is how many square units?
(Inv. 7)

6. Compare: $\frac{4}{5} \bigcirc 0.6$
(76)

7. If $\frac{1}{3}$ of the dozen eggs were cracked, how many eggs were not cracked?
(77)

8. Some triangles are quadrilaterals. True or false?
(60)

9. Sixty percent of the 30 students in the class were girls. Write 60% as a reduced fraction. Then find the number of girls in the class.
(41)

10. $5 \div 0.5$
(49)

11. $4.32 + 5 + 6.7$
(38)

12. $2.8 \div \sqrt{64}$
(38, 45)

13. $3.6 \div 10^2$
(52)

14. $\frac{2}{3} - \frac{1}{6}$
(55)

15. $4\frac{1}{2} + 2\frac{3}{4}$
(59)

16. $4\frac{1}{4} - 3\frac{3}{4}$
(63)

17. $2\frac{1}{2} \times 1\frac{1}{5}$
(70)

18. $\frac{9}{25} = \frac{?}{100}$
(42)

19. $2 \div 1\frac{2}{3}$
(68)

20. Round 0.5678 to the nearest hundredth.
(51)

TEST

16B

Also take Facts Practice Test K (Linear Measurement).

Name ______________________

1. A half gallon of milk is how many quarts?
(78)

2. Ken correctly answered 43 of the 50 questions. What percent of the questions did he answer correctly?
(75)

3. Figure *ABCD* is a rectangle. Segment *AB* is 3 cm long, segment *BC* is 4 cm long, and segment *BD* is 5 cm long. What is the area of triangle *BCD?*
(79)

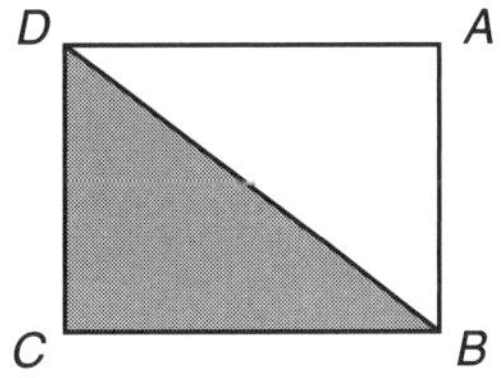

The coordinates of the four vertices of a square are (1, –1), (4, –1), (4, 2), and (1, 2). Use this information for problems 4 and 5.

4. How many units is the perimeter of the square?
(Inv. 7)

5. How many square units is the area of the square?
(Inv. 7)

6. Compare: $\frac{1}{4} \bigcirc 0.3$
(76)

7. If $\frac{1}{6}$ of the one dozen eggs were cracked, how many eggs were not cracked?
(77)

8. All squares are rectangles. True or false?
(60)

9. Keisha spent 40% of an hour on her math homework. Write 40% as a reduced fraction. Then find the number of minutes Keisha spent on her math homework.
(41)

10. $4 \div 0.4$
(49)

11. $12 + 4.87 + 6.3$
(38)

12. $3.4 \div \sqrt{25}$
(38, 45)

13. $5.2 \div 10^2$
(52)

14. $\frac{5}{6} - \frac{1}{3}$
(55)

15. $6\frac{5}{8} + 1\frac{1}{2}$
(59)

16. $6\frac{1}{2} - 2\frac{7}{8}$
(63)

17. $1\frac{1}{4} \times 2\frac{2}{5}$
(70)

18. $\frac{7}{25} = \frac{?}{100}$
(42)

19. $2 \div 2\frac{1}{3}$
(68)

20. Round 5.4638 to the nearest tenth.
(51)

TEST 17A

Also take Facts Practice Test K (Linear Measurement).

Name ____________________

1. (41) Write 80% as a reduced fraction and as a decimal numeral.

2. (79) What is the area of this triangle?

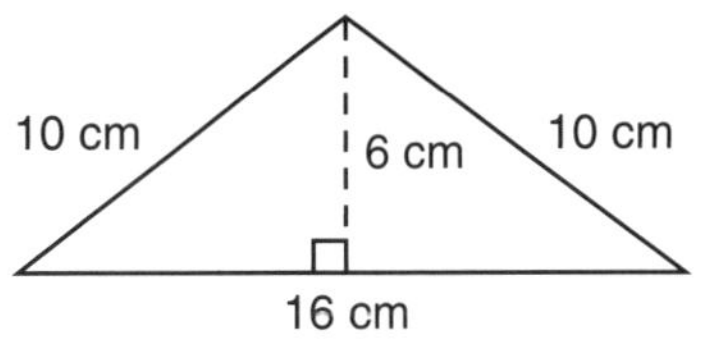

3. (75) Write $\frac{3}{5}$ as a percent.

4. (47) The diameter of a circle is 100 cm. What is the circumference of the circle? (Use 3.14 for π.)

Refer to parallelogram *ABCD* to answer problems 5–7.

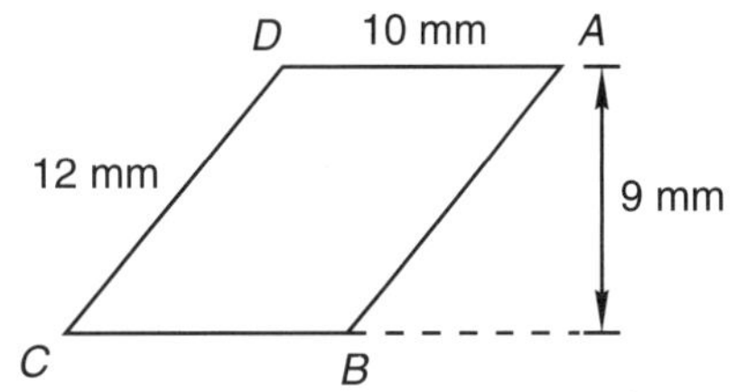

5. (71) What is the perimeter of the parallelogram?

6. (71) What is the area of the parallelogram?

7. (71) If angle *A* measures 50°, what is the measure of angle *C?*

8. (23, 29) What is the ratio of pigs to goats in a barnyard with 12 pigs and 8 goats?

9. (Inv. 7) The coordinates of three vertices of a rectangle are (3, 2), (3, –1), and (–1, –1). What are the coordinates of the fourth vertex?

10. (22) If $\frac{3}{4}$ of the 2000 raffle tickets were sold, how many tickets were sold?

11. (81) 3 ft – 10 in. (Write the difference in inches.)

12. (84) $3 \times 4 + 2 \times 5$

13. (82) What is the volume of a shoe box that is 12 inches long, 6 inches wide, and 5 inches high?

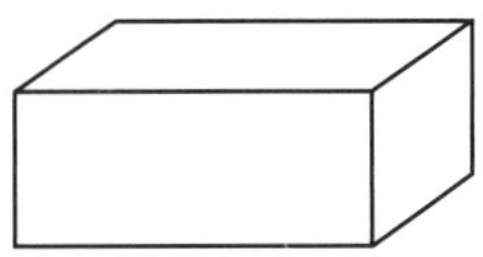

14. (85) Complete this proportion: $\frac{6}{9} = \frac{10}{w}$

15. (40) $10 - (0.1 + 1)$

16. (87) Solve: $3n = 0.45$

17. (59) $3\frac{3}{4} + 1\frac{5}{8}$

18. (63) $2\frac{1}{2} - \frac{3}{4}$

19. (70) $2\frac{1}{4} \times 2$

20. (68) $5 \div 3\frac{3}{4}$

TEST 17B

Also take Facts Practice Test K (Linear Measurement).

Name ______________________

1. (41) Write 60% as a reduced fraction and as a decimal numeral.

2. (79) What is the area of this triangle?

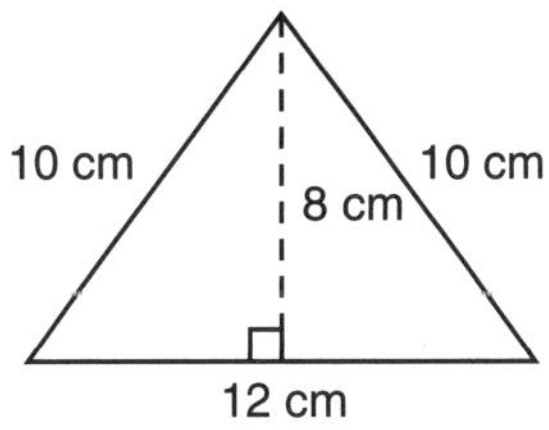

3. (75) Write $\frac{4}{5}$ as a percent.

4. (47) The diameter of a circle is 200 cm. What is the circumference of the circle? (Use 3.14 for π.)

Refer to parallelogram *ABCD* to answer problems 5–7.

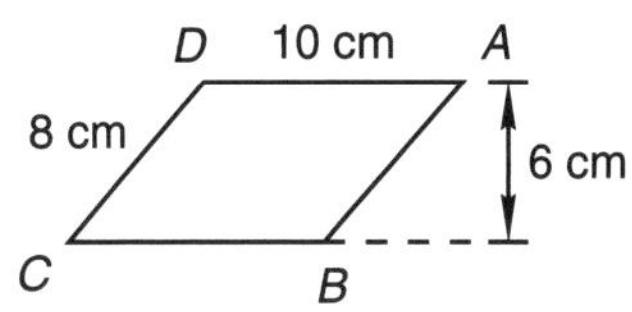

5. (71) What is the perimeter of the parallelogram?

6. (71) What is the area of the parallelogram?

7. (71) If angle *B* measures 130°, what is the measure of angle *C?*

8. (23, 29) What is the ratio of children to adults in a theater with 80 children and 60 adults?

9. (Inv. 7) The coordinates of three vertices of a rectangle are (4, 3), (4, 0), and (–2, 0). What are the coordinates of the fourth vertex?

10. (22) If Sarah answered $\frac{4}{5}$ of the 20 questions correctly, how many questions did she answer correctly?

11. (81) 3 ft – 6 in. (Write the difference in inches.)

12. (84) $3 \times 4 - 2 \times 5$

13. (82) What is the volume of a tissue box that is 10 inches long, 4 inches wide, and 5 inches high?

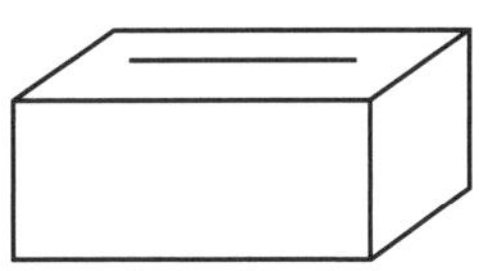

14. (85) Complete this proportion: $\frac{9}{m} = \frac{6}{10}$

15. (40) $2 - (0.2 + 0.02)$

16. (87) Solve: $4n = 0.56$

17. (59) $1\frac{3}{4} + 1\frac{1}{2}$

18. (63) $4\frac{1}{2} - 1\frac{3}{4}$

19. (70) $5 \times 1\frac{3}{5}$

20. (68) $5 \div 6\frac{2}{3}$

TEST

18A

Also take Facts Practice Test L (Liquid Measurement).

Name ______________________

1. (77) If $\frac{3}{4}$ of the 200 seeds sprouted, how many seeds did not sprout?

2. (88) The ratio of boys to girls in the class was 3 to 2. If there were 12 boys, how many girls were there?

Refer to triangle *ABC* to answer problems 3 and 4.

3. (8) What is the perimeter of the triangle?

4. (79) What is the area of the triangle?

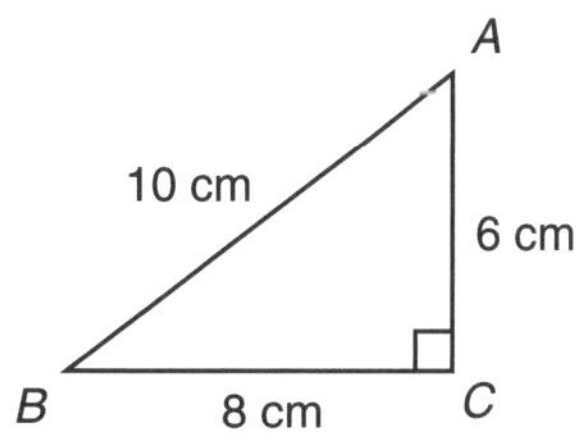

5. (40) $5 - 3.21$

6. (39) 2.7×0.03

7. (38, 45) $0.27 \div \sqrt{36}$

8. (49) $27 \div 0.9$

9. (89) Between which two numbers is $\sqrt{20}$?

A. 2 and 3 B. 3 and 4 C. 4 and 5 D. 19 and 21

10. (55) $\frac{5}{6} + \frac{1}{2}$

11. (63) $3\frac{1}{3} - 2\frac{1}{2}$

12. (70) $3\frac{3}{4} \times 1\frac{3}{5}$

13. (68) $5 \div 1\frac{1}{4}$

14. (87) Solve: $5y = 9$

Refer to the circle on the right to answer problems 15 and 16. (Use 3.14 for π.)

15. (47) What is the circumference of the circle?

16. (86) What is the area of the circle?

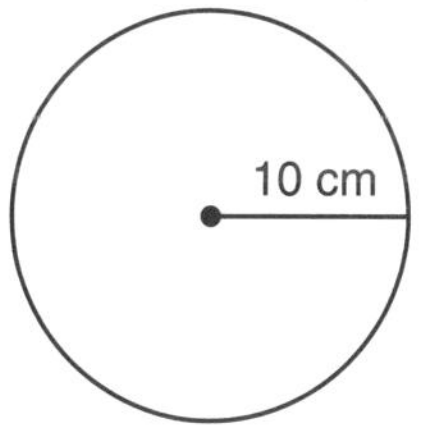

17. (41) Write 75% as a reduced fraction and as a decimal numeral.

18. (82) What is the volume of this rectangular prism?

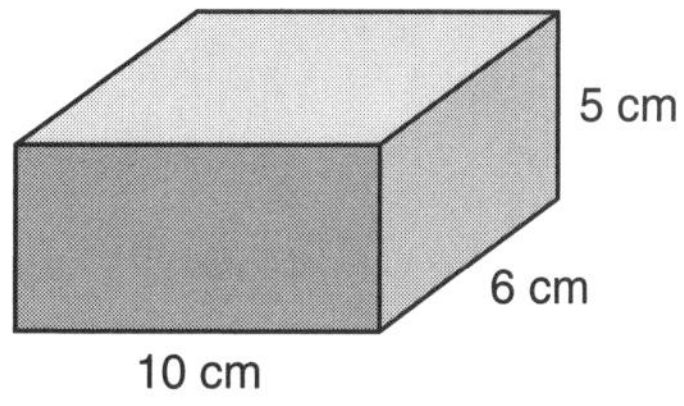

19. (Inv. 6) A cube has how many faces?

20. (Inv. 7) The coordinates of three vertices of a parallelogram are (3, –2), (5, 2), and (0, 2). What are the coordinates of the fourth vertex?

TEST 18B

Also take Facts Practice Test L (Liquid Measurement).

Name ______________________

1. (77) If $\frac{4}{5}$ of the 200 seeds sprouted, how many seeds did not sprout?

2. (88) The ratio of boys to girls in the class was 3 to 2. If there were 12 girls, how many boys were there?

Refer to triangle *ABC* to answer problems 3 and 4.

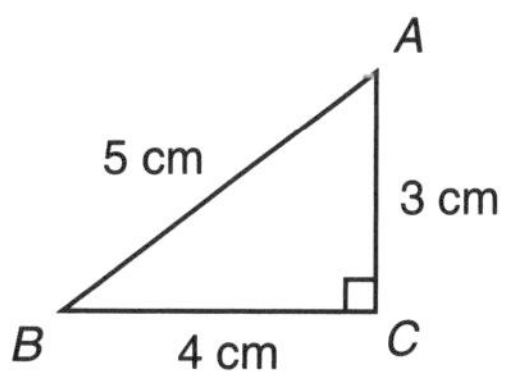

3. (8) What is the perimeter of the triangle?

4. (79) What is the area of the triangle?

5. (40) $6 - 4.3$

6. (39) 4.3×0.02

7. (38, 45) $0.45 \div \sqrt{36}$

8. (49) $45 \div 0.9$

9. (89) Between which two numbers is $\sqrt{30}$?

A. 3 and 4 B. 4 and 5 C. 5 and 6 D. 29 and 31

10. (55) $\frac{2}{3} + \frac{5}{6}$

11. (63) $3\frac{1}{4} - 2\frac{1}{2}$

12. (70) $1\frac{1}{5} \times 1\frac{2}{3}$

13. (68) $5 \div 1\frac{2}{3}$

14. (87) Solve: $5y = 8$

Refer to the circle on the right to answer problems 15 and 16. (Use 3.14 for π.)

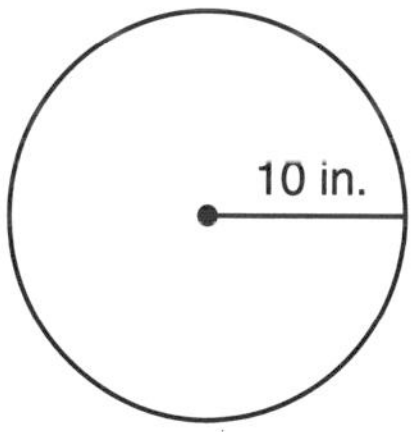

15. (47) What is the circumference of the circle?

16. (86) What is the area of the circle?

17. (41) Write 65% as a reduced fraction and as a decimal numeral.

18. (82) What is the volume of this rectangular prism?

19. (Inv. 6) A cube has how many edges?

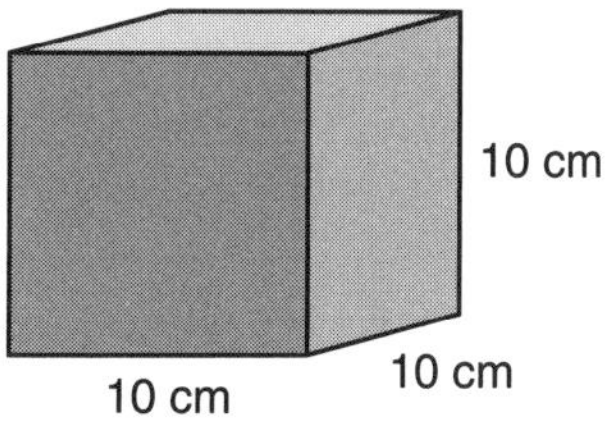

20. (Inv. 7) The coordinates of three vertices of a parallelogram are (3, 3), (0, 3), and (–2, 0). What are the coordinates of the fourth vertex?

TEST 19A

Also take Facts Practice Test L (Liquid Measurement).

Name ______________________

1. (58) What is the probability of rolling an even number with one roll of a number cube?

2. (Inv. 5) Margie's first seven test scores are shown below. What is the median of these scores?

90, 85, 90, 85, 100, 85, 100

3. (6, 75) What percent of this circle is shaded?

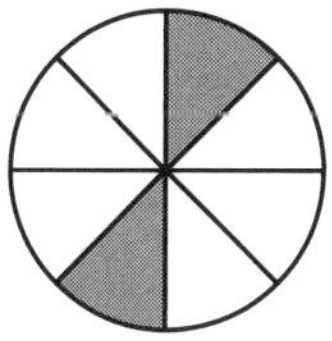

4. (Inv. 6) A pyramid with a square base has how many edges?

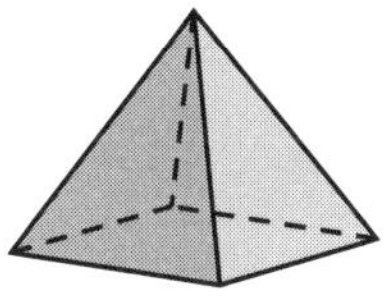

5. (79) What is the area of triangle *ABC?*

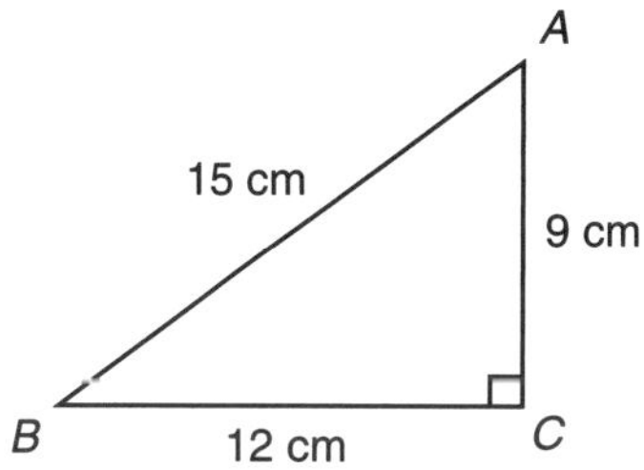

6. (93) Which of these terms describes triangle *ABC?*

A. acute triangle B. equilateral triangle C. right triangle D. isosceles triangle

7. (92) $10 + 5 \times 4 - 2^2$

8. (92) $\left(1\frac{1}{2}\right)^2$

9. (85) Solve: $\frac{4}{6} = \frac{10}{w}$

10. (92) $4^2 - \sqrt{16}$

11. (5) $\frac{(15)(16)}{20}$

12. (22) What number is $\frac{5}{8}$ of 56?

13. (40) $6 - (0.37 + 1.2)$

14. (49) $4 \div 0.5$

15. (59) $6\frac{2}{3} + 3 + 4\frac{1}{2}$

16. (63) $3\frac{1}{3} - 2\frac{3}{4}$

17. (68) $1\frac{4}{5} \div 3$

18. (94) To change a fraction to a percent, multiply the fraction by 100%. Change $\frac{1}{3}$ to a percent.

19. (37, 74) Write $3\frac{3}{4}$ as a decimal numeral. Then add it to 1.5 and find the sum.

20. (Inv. 7) How many units is the perimeter of a rectangle with vertices located at (–2, 2), (–2, –2), (4, –2), and (4, 2)?

TEST

19B

Also take Facts Practice Test L (Liquid Measurement).

Name ____________________

1. (58) What is the probability of rolling an odd number with one roll of a number cube?

2. (Inv. 5) Ibrahim's first seven quiz scores are shown below. What is the median of these scores?

17, 19, 18, 20, 19, 18, 20

3. (6, 75) What percent of this circle is shaded?

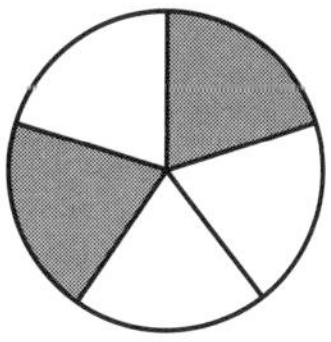

4. (Inv. 6) A triangular prism has how many faces?

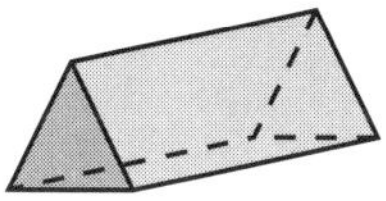

5. (79) What is the area of triangle *ABC?*

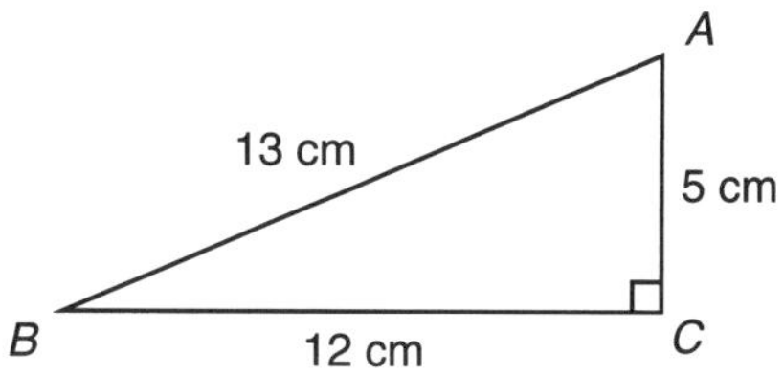

6. (93) Which of these terms describes triangle *ABC?*

A. equilateral triangle B. isosceles triangle C. obtuse triangle D. right triangle

7. (92) $8 + 4 \times 2 - 3^2$

8. (92) $\left(2\frac{1}{2}\right)^2$

9. (85) Solve: $\frac{6}{9} = \frac{8}{n}$

10. (92) $3^3 - \sqrt{9}$

11. (5) $\frac{(25)(18)}{15}$

12. (22) What number is $\frac{3}{5}$ of 60?

13. (40) $6 + (1.2 - 0.37)$

14. (49) $3 \div 0.5$

15. (59) $3\frac{3}{8} + 4 + 2\frac{3}{4}$

16. (63) $3\frac{2}{3} - 2\frac{3}{4}$

17. (68) $5\frac{1}{3} \div 4$

18. (94) To change a fraction to a percent, multiply the fraction by 100%. Change $\frac{2}{3}$ to a percent.

19. (7, 74) Write $3\frac{3}{4}$ as a decimal numeral. Then add it to 2.5 and find the sum.

20. (Inv. 7) The coordinates of the vertices of a rectangle are (5, 4), (0, 4), (0, 0), and (5, 0). The perimeter of the rectangle is how many units?

TEST

20A

Also take Facts Practice Test M (24 Percent-Fraction-Decimal Equivalents).

Name ______________________

1. (88) The ratio of A's to B's on the last test was 3 to 4. If there were 12 A's, how many B's were there?

2. (Inv. 10) What is the probability of rolling a 4 twice with two rolls of a number cube?

Complete this table.

	FRACTION	DECIMAL	PERCENT
3. (99)	(a)	0.6	(b)
4. (99)	(a)	(b)	5%

5. (77) Three fifths of the 30 students finished. How many did not finish?

Refer to the parallelogram at the right to answer problems 6 and 7.

6. (71) What is the area of the parallelogram?

7. (71) What is the perimeter of the parallelogram?

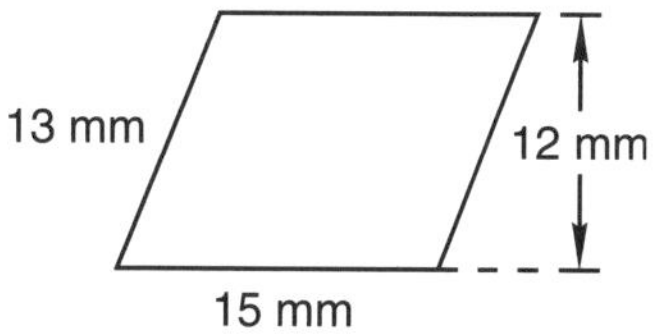

8. (97) Parallel lines m and n are intersected by line p as shown. Angle 2 measures 80°. What is the measure of $\angle 6$?

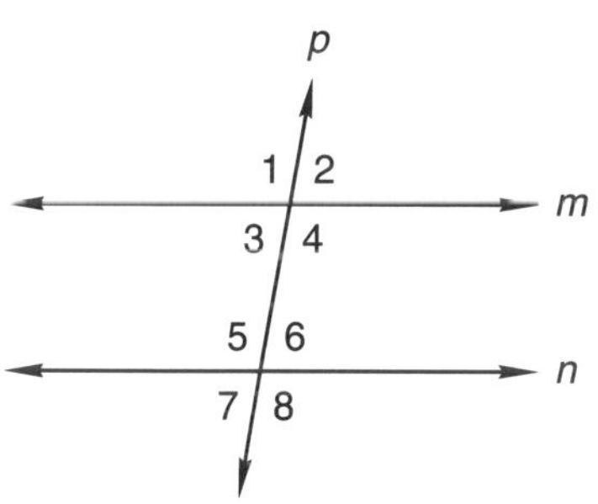

9. (90) Anita began at point A and ran around the field once making the five turns shown. In those five turns Anita turned a total of how many degrees?

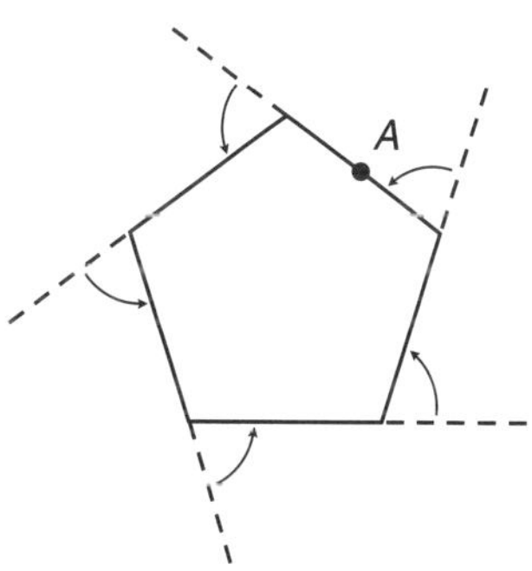

10. (89) $\sqrt{900}$

11. (100) $-3 + -5$

12. (100) $-5 - (-7)$

13. (82) What is the volume of a box that is 5 cm long, 4 cm wide, and 3 cm high?

14. (Inv. 6) A cube has how many more edges than it has vertices?

15. (92) $2^3 - \sqrt{9} + 3 \times 2$

16. (74) $6.25 - 4\frac{1}{2}$ (decimal answer)

17. (85) Solve: $\frac{3}{20} = \frac{m}{100}$

18. (43) Solve: $y + 3.6 = 5$

19. (100) At 6 a.m. the temperature was −6°F. By noon the temperature had risen to 12°F. How many degrees had the temperature risen?

20. (86) The diameter of a circular target is 12 inches. What is the area of the target? (Use 3.14 for π.)

TEST

20B

Also take Facts Practice Test M (24 Percent-Fraction-Decimal Equivalents).

Name ______________________

1. (88) The ratio of A's to B's on the last test was 4 to 3. If there were 12 A's, how many B's were there?

2. (Inv. 10) What is the probability of rolling a 3 twice with two rolls of a number cube?

Complete this table.

	FRACTION	DECIMAL	PERCENT
3. (99)	(a)	0.4	(b)
4. (99)	(a)	(b)	4%

5. (77) Two thirds of the 30 students finished. How many did not finish?

Refer to the parallelogram at the right to answer problems 6 and 7.

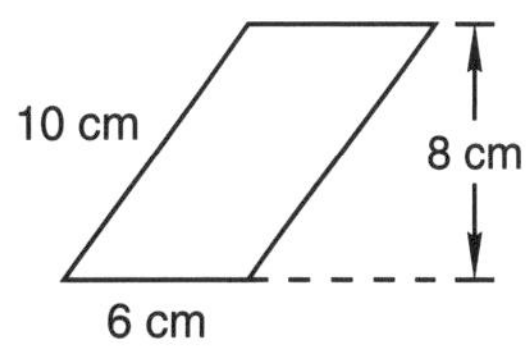

6. (71) What is the area of the parallelogram?

7. (71) What is the perimeter of the parallelogram?

8. (97) Parallel lines m and n are intersected by line p as shown. Angle 1 measures 105°. What is the measure of $\angle 5$?

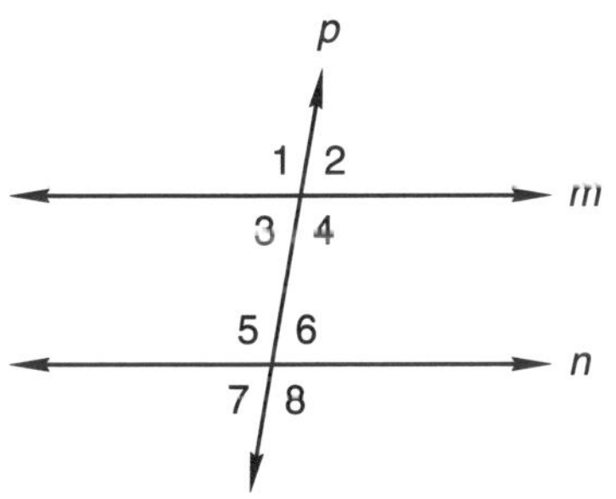

9. (90) Andy began at point A and ran around the field once making the six turns shown. In those six turns Andy turned a total of how many degrees?

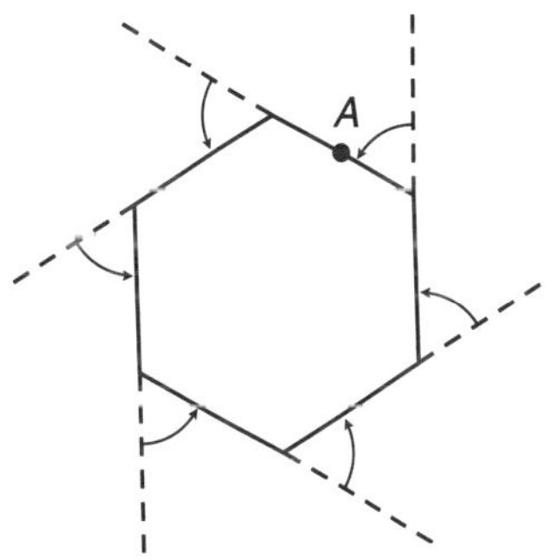

10. (89) $\sqrt{400}$

11. (100) $-2 + -5$

12. (100) $-3 - (-7)$

13. (82) What is the volume of a cube that has edges 3 cm long?

14. (Inv. 6) A cube has how many more vertices than faces?

15. (92) $2^3 + \sqrt{9} - 3 \times 2$

16. (74) $6\frac{1}{2} + 4.25$ (decimal answer)

17. (85) Solve: $\frac{3}{25} = \frac{m}{100}$

18. (43) Solve: $y + 2.7 = 5$

19. (100) At 6 a.m. the temperature was –5°F. By noon the temperature had risen to 8°F. How many degrees had the temperature risen?

20. (86) The diameter of a circular target is 16 inches. What is the area of the target? (Use 3.14 for π.)

TEST 21A

Also take Facts Practice Test M (24 Percent-Fraction-Decimal Equivalents).

Name ______________________

1. Divide $1.30 by 8 and round the quotient to the nearest cent.
(45, 51)

2. Forty percent of the students who took the test earned an A. If ten students earned an A, then how many students took the test? (Use a ratio box.)
(105)

3. What are the coordinates of the point halfway between (–2, –2) and (4, –2)?
(Inv. 7)

4. $10^2 - \sqrt{81} - 3 \times 10$
(92)

5. How many 1-inch sugar cubes would be needed to form this rectangular prism?
(82)

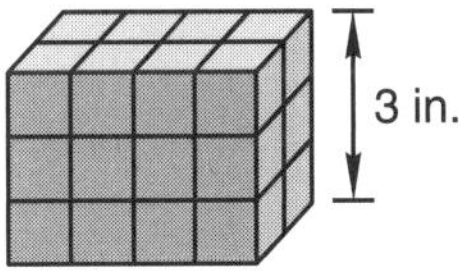

6. What is the perimeter of this figure?
(103)

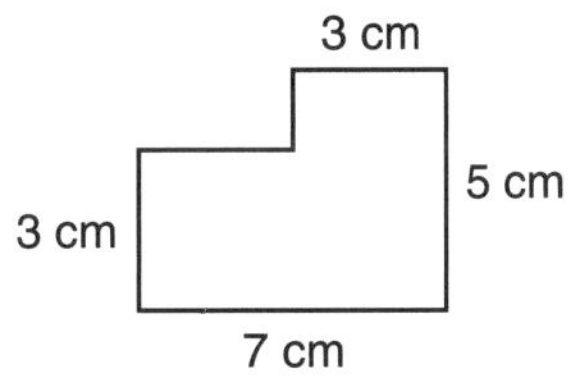

7. If the radius of a circle is 10 cm, what is its circumference? (Use 3.14 for π.)
(47)

8. Two tons is how many pounds?
(102)

9. Solve this proportion: $\frac{6}{10} = \frac{a}{25}$
(85)

10. The boy-girl ratio in the class was 5 to 4. If there were 27 students, how many boys were there? (Draw a ratio box for the problem. Then write and solve a proportion to find the answer.)
(101)

Complete this table.

	FRACTION	DECIMAL	PERCENT
11. *(99)*	$1\frac{3}{5}$	(a)	(b)
12. *(99)*	(a)	0.9	(b)
13. *(99)*	(a)	(b)	10%

14. $6.3 - (1.25 + 2)$
(40)

15. $(-3) + (-2) - (-4)$
(104)

16. $4 \div (0.2 \times 2)$
(49)

17. $6\frac{2}{3} + 3\frac{3}{4}$
(59)

18. $6\frac{1}{4} \times 1\frac{3}{5}$
(70)

19. $2\frac{1}{3} \div 1\frac{5}{9}$
(68)

20. What is the area of the shaded part of this parallelogram?
(79)

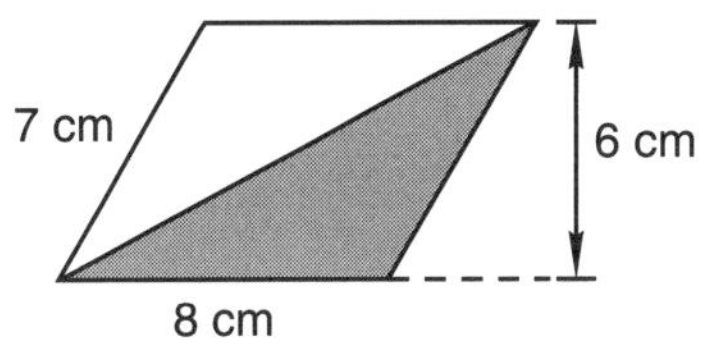

TEST

21B

Also take Facts Practice Test M (24 Percent-Fraction-Decimal Equivalents).

Name ______________________

1. Divide \$6.50 by 8 and round the quotient to the nearest cent.
5, 51)

2. Thirty percent of the students who took the test earned an A. If 9 students earned an A, then how many students took the test? (Use a ratio box.)
(105)

3. What are the coordinates of the point halfway between (–2, 3) and (6, 3)?
nv. 7)

4. $10^2 + \sqrt{9} + 3 \times 10$
(92)

5. How many 1-centimeter sugar cubes would be needed to form this cube?
(82)

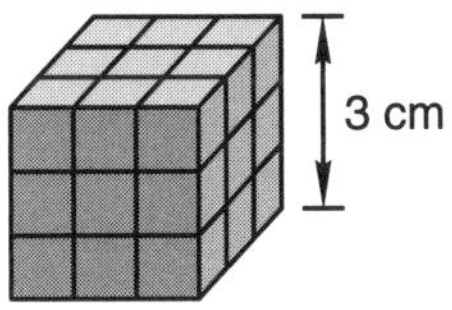

6. What is the perimeter of this figure?
(103)

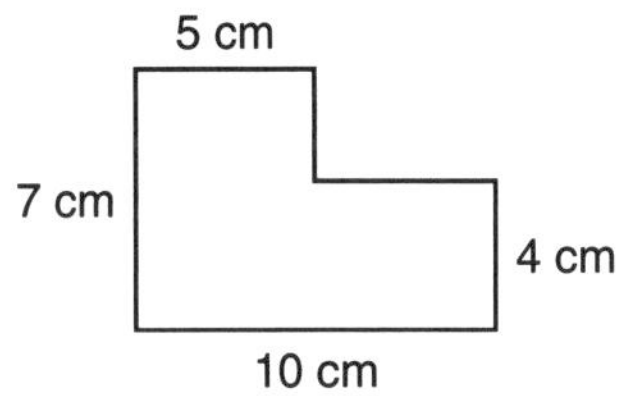

7. If the radius of a circle is 10 cm, what is its area? (Use 3.14 for π.)
(86)

8. Half of a ton is how many pounds?
(102)

9. Solve this proportion: $\frac{8}{12} = \frac{12}{w}$
(85)

10. The ratio of boys to girls was 4 to 5. If there were 27 students, how many boys were there? (Draw a ratio box for the problem. Then write and solve a proportion to find the answer.)
(101)

Complete this table.

	FRACTION	DECIMAL	PERCENT
11. (99)	$2\frac{1}{5}$	(a)	(b)
12. (99)	(a)	0.7	(b)
13. (99)	(a)	(b)	5%

14. $4.2 + (3 - 1.75)$
(40)

15. $(-2) + (-3) - (-5)$
(104)

16. $0.12 \times (0.12 \div 6)$
(49)

17. $1\frac{3}{4} + 1\frac{2}{3}$
(59)

18. $8\frac{1}{3} \times 1\frac{4}{5}$
(70)

19. $2\frac{2}{5} \div 1\frac{4}{5}$
(68)

20. What is the area of the shaded part of this parallelogram?
(79)

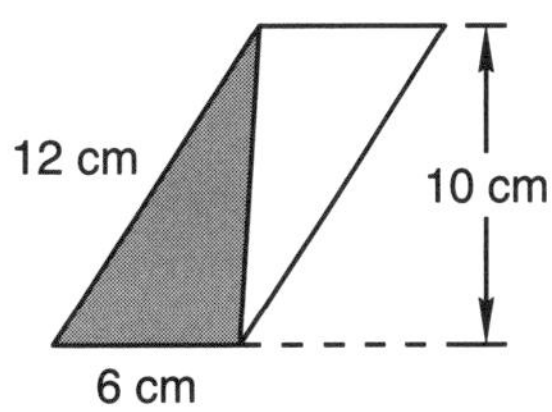

TEST

22A

Also take Facts Practice Test M (24 Percent-Fraction-Decimal Equivalents).

Name ______________________

1. (98) Triangle ABC is an isosceles triangle. Angles B and C each measure 70°. What is the measure of angle A?

2. (110) Trace triangle ABC on your paper and draw its line of symmetry.

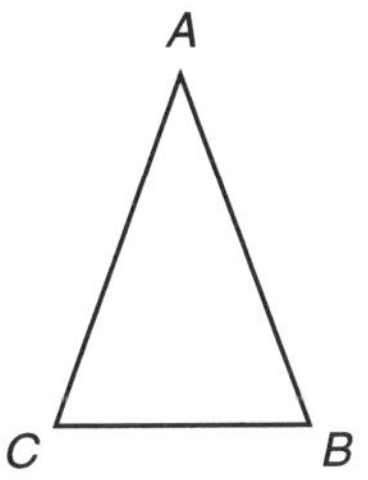

3. (63) $4\frac{1}{2} - 1\frac{5}{6}$

4. (68) $6\frac{2}{3} \div 1\frac{1}{4}$

5. (37, 74) $3.75 + \frac{3}{8}$ (decimal answer)

6. (14, 99) Arrange in order from least to greatest: $\frac{1}{2}$, −2, 0.2

7. (Inv. 7, 79) The coordinates of the vertices of a triangle are (0, 0), (6, 0), and (3, 4). What is the area of the triangle?

8. (104) $(-6) + (-4) - (3)$

9. (Inv. 10) A bag contains 3 red, 2 white, and 5 blue marbles. If one marble is drawn and not replaced and another marble is drawn, what is the probability that both marbles will be red?

10. (92) $3^3 - 2^3 - 2 \times 3$

Complete this table.

	FRACTION	DECIMAL	PERCENT
11. (99)	$1\frac{1}{5}$	(a)	(b)
12. (99)	(a)	0.3	(b)
13. (99)	(a)	(b)	35%

14. (82) What is the volume of this rectangular prism?

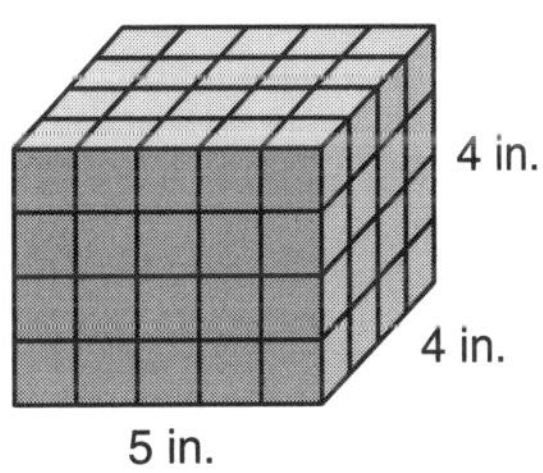

15. (86) The diameter of a circular mirror is 20 inches. What is the area of the mirror? (Use 3.14 for π.)

16. (101) What is the boy-girl ratio in a class of 32 pupils if there are 18 girls?

17. (85) Solve the proportion: $\frac{a}{12} = \frac{20}{30}$

18. (106) Solve: $3n + 2 = 35$

Refer to the hexagon on the right to answer problems 19 and 20.

19. (103) What is the perimeter of the hexagon?

20. (107) What is the area of the hexagon?

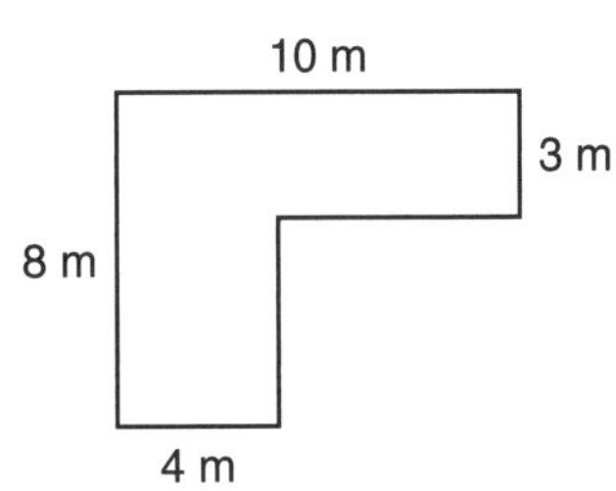

TEST

22B

Also take Facts Practice Test M (24 Percent-Fraction-Decimal Equivalents).

Name ______________________

1. (98) Triangle *ABC* is an isosceles triangle. Angles *B* and *C* each measure 75°. What is the measure of $\angle A$?

2. (110) Trace triangle *ABC* on your paper and draw its line of symmetry.

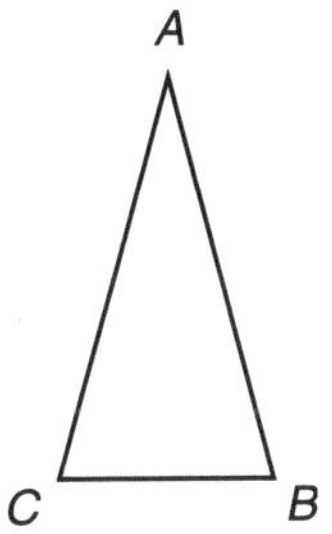

3. (63) $5\frac{1}{3} - 3\frac{1}{2}$

4. (68) $3\frac{3}{4} \div 1\frac{1}{4}$

5. 37, 74) $4.25 + \frac{3}{8}$ (decimal answer)

6. (14, 99) Arrange in order from least to greatest: $\frac{3}{4}$, 0.35, –3

7. 7, 79) The coordinates of the vertices of a triangle are (1, 1), (7, 1), and (5, 5). What is the area of the triangle?

8. (104) $(-7) + (-2) - (3)$

9. nv. 10) A bag contains 3 red, 2 white, and 5 blue marbles. If one marble is drawn and not replaced and another marble is drawn, what is the probability that both marbles will be blue?

10. (92) $3^2 - 2^3 + 2 \times 3$

Complete this table.

	FRACTION	DECIMAL	PERCENT
11. (99)	$1\frac{1}{4}$	(a)	(b)
12. (99)	(a)	0.7	(b)
13. (99)	(a)	(b)	15%

14. (82) The rectangular prism below is made up of how many small blocks?

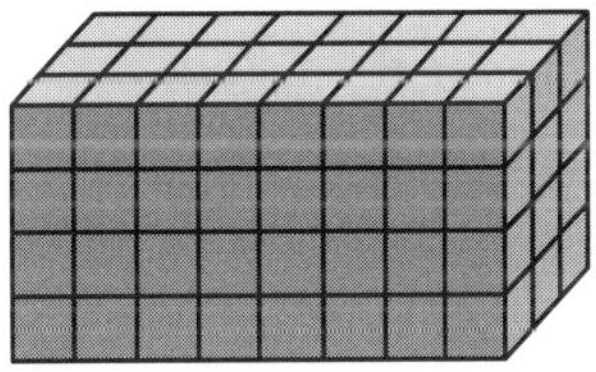

15. (47) The radius of John's bicycle wheel is 16 inches. What is the circumference of the wheel? (Use 3.14 for π.)

16. (101) What is the boy-girl ratio in a class of 32 pupils if there are 18 boys?

17. (85) Solve the proportion: $\frac{10}{a} = \frac{25}{30}$

18. (106) Solve: $4m + 4 = 40$

Refer to the hexagon on the right to answer problems 19 and 20.

19. (103) What is the perimeter of the hexagon?

20. (107) What is the area of the hexagon?

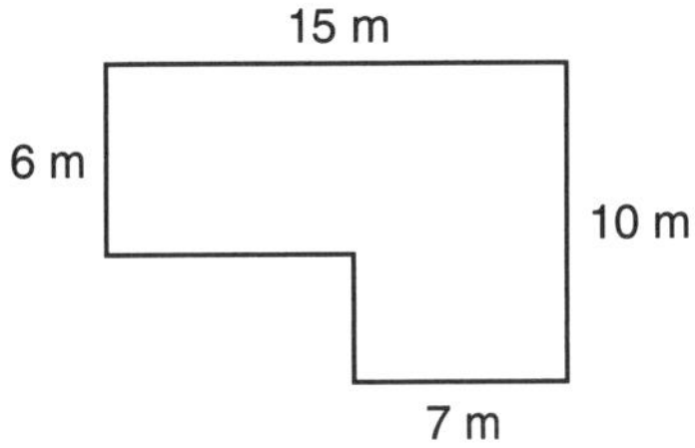

TEST

23A

Also take Facts Practice Test N (Measurement Facts).

Name ________________________

1. (111) One hundred ten students are assigned to four classrooms as equally as possible. How many students are in each classroom? (Write four numbers in all to show how many students are in each room.)

2. (113) Write $1\frac{1}{2}$ million as a standard numeral.

3. (73) Write the prime factorization of 81 using exponents.

4. (114) Show how to convert 8 quarts to gallons using a unit multiplier. (1 gal = 4 qt)

5. (92) Write in standard notation: $(5 \times 10^3) + (4 \times 10^1)$

6. (. 10) In a bag are three red marbles and three green marbles. If two marbles are drawn from the bag at the same time, what is the probability that both marbles will be red?

7. (93) What is the perimeter of an equilateral triangle if one of its sides is 15 centimeters long?

8. (109) The two triangles at right are congruent. What is the perimeter of each?

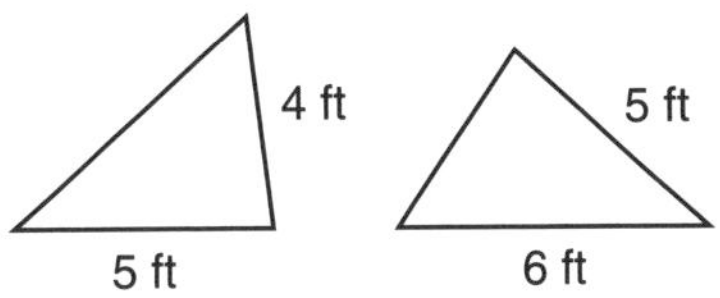

9. (47) If the diameter of a wheel is 10 inches, what is the circumference of the wheel? (Use 3.14 for π.)

Refer to the figure on the right for problems 10 and 11.

10. (103) What is the perimeter of the hexagon?

11. (107) What is the area of the hexagon?

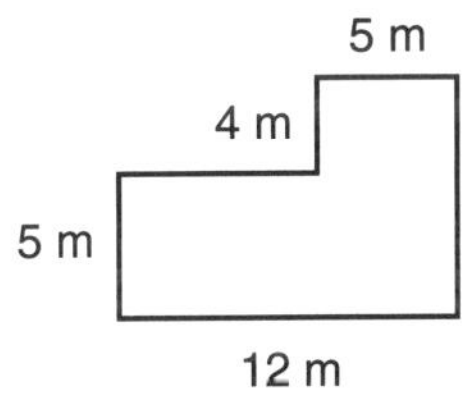

Complete this table.

	FRACTION	DECIMAL	PERCENT
12. (99)	$\frac{3}{4}$	(a)	(b)
13. (99)	(a)	0.02	(b)
14. (99)	(a)	(b)	120%

15. (5, 38) $0.475 + (5 - 0.45)$

16. (49) $0.6 \div (0.4 \times 0.5)$

17. (112) $(-4)(-7)$

18. (112) $\dfrac{-16}{+2}$

19. (79) The area of the shaded part of this rectangle is half of the area of the rectangle. What is the area of the shaded triangle?

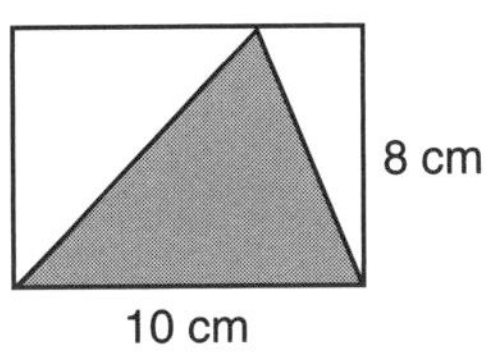

20. (105) Julie missed 5 questions on the test but answered 80% of the questions correctly. How many questions were on the test?

TEST

23B

Also take Facts Practice Test N (Measurement Facts).

Name ______________________

1. (111) Eighty-five students are assigned to three classrooms as equally as possible. How many students are assigned to each classroom? (Write three numbers in all to show how many students are in each room.)

2. (113) Write 1.5 billion as a standard numeral.

3. (73) Write the prime factorization of 64 using exponents.

4. (114) Show how to convert 8 quarts to pints using a unit multiplier. (1 qt = 2 pt)

5. (92) Write in standard notation: $(6 \times 10^2) + (4 \times 10^1)$

6. (Inv. 10) In a bag are three red marbles and three green marbles. If two marbles are drawn from the bag at the same time, what is the probability that both marbles will be green?

7. (93) If the perimeter of an equilateral triangle is 36 inches, how long is each side?

8. (109) If two squares are congruent and if the perimeter of one is 20 cm, what is the area of the other?

9. (47) If the radius of a wheel is 10 inches, what is the circumference of the wheel? (Use 3.14 for π.)

Refer to the hexagon on the right for problems 10 and 11.

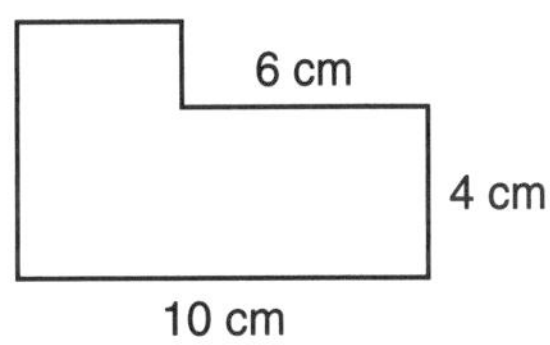

10. (103) What is the perimeter of the hexagon?

11. (107) What is the area of the hexagon?

Complete this table.

	FRACTION	DECIMAL	PERCENT
12. (99)	$\frac{1}{4}$	(a)	(b)
13. (99)	(a)	0.04	(b)
14. (99)	(a)	(b)	150%

15. (5, 38) $6.23 + (4 - 0.5)$

16. (49) $6 \div (0.4 \times 0.5)$

17. (112) $(-9)(-4)$

18. (112) $\frac{-18}{+3}$

19. (79) The area of the shaded part of this rectangle is half of the area of the rectangle. What is the area of the shaded triangle?

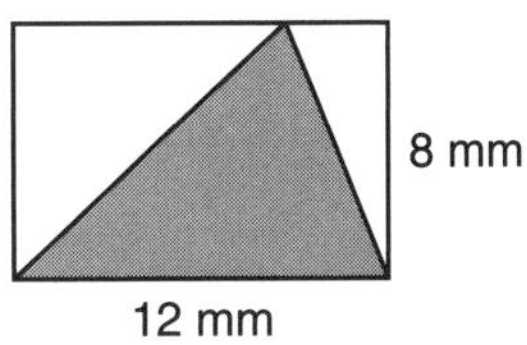

20. (105) James missed 6 questions on the test but answered 80% of the questions correctly. How many questions were on the test?

Test-Day Activity Masters

Test day can be structured so that students complete a facts practice, a test, and a brief instructional activity. Therefore, instructional activities called Test-Day Activities are included in each Cumulative Assessment section in the *Saxon Math 7/6 Teacher's Manual.* Masters for use with the Test-Day Activities are on the following pages. After each test distribute one copy of the suggested Test-Day Activity Master(s) to each student, and conduct the activity using the instruction in the Teacher's Manual. To check students' work, refer to the Test-Day Activity Masters Answers section in the back of this book.

Note: To provide maximum time for the test and Test-Day Activity, we recommend that no homework be collected or reviewed on test day.

1 Patterns

For use with Test-Day Activity 1

Name ______________________

sequence A list of numbers arranged according to a certain rule.

arithmetic sequence A sequence in which each term is found by adding a fixed amount to the previous term. For example:

5, 9, 13, 17, 21, . . .

geometric sequence A sequence in which each term is found by multiplying the previous term by a fixed amount. For example:

2, 6, 18, 54, 162, . . .

2 Patterns

For use with Test-Day Activity 1

Name ____________________

1. 3, 6, 12, 24, . . .

a. Is this an arithmetic sequence or a geometric sequence?

b. In words, describe the pattern.

c. Find the next three terms in the sequence.

2. 13, 17, 21, 25, 29, . . .

a. Is this an arithmetic sequence or a geometric sequence?

b. In words, describe the pattern.

c. Find the 6th term, 7th term, and 8th term in the sequence.

3. This arithmetic sequence counts down. Find the next two terms.

100, 94, 88, 82, ____, ____

Not all sequences are arithmetic sequences and geometric sequences. Find a rule for the sequence or pattern, and use it to write the next two terms.

4. 1, 2, 4, 7, 11, 16, ____, ____

5. AB, DE, GH, JK, ____, ____

3

Patterns

For use with Test-Day Activity 1

Name ______________________

Find the missing terms in the sequence or pattern.

1. 10, 15, 20, 25, 30, ____, ____, ____, . . .

2. 1, 4, 9, 16, 25, 36, ____, ____, ____ , . . .

3. 10, 15, 21, 28, 36, ____, ____, ____, . . .

4. M, T, W, T, ____, ____, ____, . . .

5. 11, 22, ____ ,____, 55, ____, . . .

6. ____, ____, ____, 16, 20, 24, 28, . . .

7. Give an example of a geometric sequence.

8. Give an example of an arithmetic sequence.

9. In words, describe this pattern. 30, 37, 44, 51, 58, . . .

10. In words, describe this pattern. 1, 1, 2, 3, 5, 8, 13, . . .

TEST-DAY ACTIVITY MASTER

4

Make This Number

For use with Test-Day Activity 3

Name ______________________________

Write an expression that uses each of the three numbers exactly once, along with any number of operations ($+$, $-$, $\times$, and $\div$) and parentheses (if desired). The expression you write must equal the given total. Problem 1 is done for you. (Hint: more than one right answer is possible.)

1. Numbers: 3, 4, 12 Total: 13
Sample answer: $12 - 3 + 4 = 13$

2. Numbers: 1, 3, 5 Total: 15

3. Numbers: 9, 9, 10 Total: 10

4. Numbers: 3, 4, 5 Total: 17

5. Numbers: 10, 11, 12 Total: 9

6. Numbers: 2, 6, 24 Total: 2

7. Numbers: 0, 6, 7 Total:42

5

Negative Numbers in the Real World

For use with Test-Day Activity 4

Name ______________________

Discuss whether a positive or negative number should be used to represent each situation.

1. $75 was deposited into your bank account.

2. The world's largest subterranean lake is located in Bonne Terre, Missouri. It lies 150 feet below the surface of the earth.

3. On the first play of the Bulldogs' football game, Trey lost 7 yards.

4. As the airplane approached the airport, its rate of descent was 5 miles per hour.

5. Hot air balloons float about 700 feet above the earth.

6. On the first day of January, it was 20 degrees below zero in Fairbanks, Alaska.

TEST-DAY ACTIVITY MASTER

6

Negative Numbers in the Real World

For use with Test-Day Activity 4

Name ______________________

Write a positive or negative number to represent each situation.

1. Mr. Whitman withdrew $50 from his bank account.

2. The fuel leak was found 3 seconds before the rocket was scheduled for blast-off.

3. The tallest mountain in North America is 20,320 feet above sea level.

4. Earl's golf score was 4 strokes under par.

5. The baby gained 16 pounds in his first year.

6. The average depth of the Channel Tunnel that connects Great Britain and France is 150 feet below the surface of the water.

Write a situation that can be represented by the negative number.

7. $-10°$

8. -14 yards

9. -200 dollars

10. -164 feet

TEST-DAY ACTIVITY MASTER

7

Geometry Search

For use with Test-Day Activity 5

Name ____________________

Complete the table below by locating and identifying the geometric shapes and angles in your classroom. Make a sketch of each shape you find.

Shape	Where is it?	What does it look like?
Rectangle	Example: the top of the teacher's desk	Example: 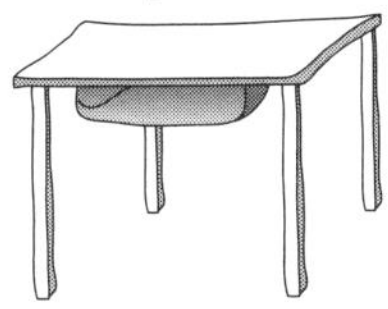
Square		
Triangle		
Circle		
Right angle		
Acute angle		
Obtuse angle		

TEST-DAY ACTIVITY MASTER

8 Constructing a Cube and Finding its Surface Area

Name ________________________

For use with Test-Day Activity 6

In this activity, you will make a cube by drawing a pattern for a cube, cutting it out, folding its edges, and taping the pattern closed.

Instructions for Constructing the Cube

a) Each of the arrangements below is a pattern for a cube. Choose one of the patterns.

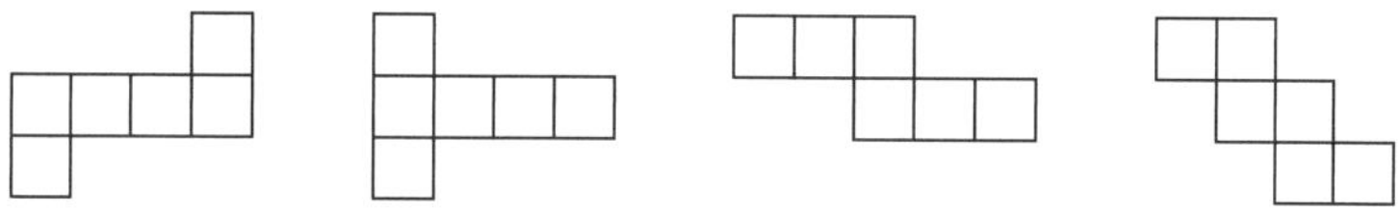

b) Cut out the dashed square on Master 9. Use it as a template for the pattern you chose.

c) Use your template to draw your pattern on cardstock or construction paper. Be sure to mark the fold lines. Cut your pattern out. Take care to cut on the outside edge only. When you are finished cutting, you should have only one piece.

d) Use your ruler to help you find the surface area of your pattern. Write the surface area directly on the pattern. (Recall that area measurements are in square units, such as square inches.)

e) Now fold the edges and tape your pattern to make a cube.

- What is the surface area of your cube? How do you know?

9

Constructing a Cube and Finding its Surface Area

For use with Test-Day Activity 6

Name ______________________

Template

10

Tangrams

For use with Test-Day Activity 7

Name ______________________

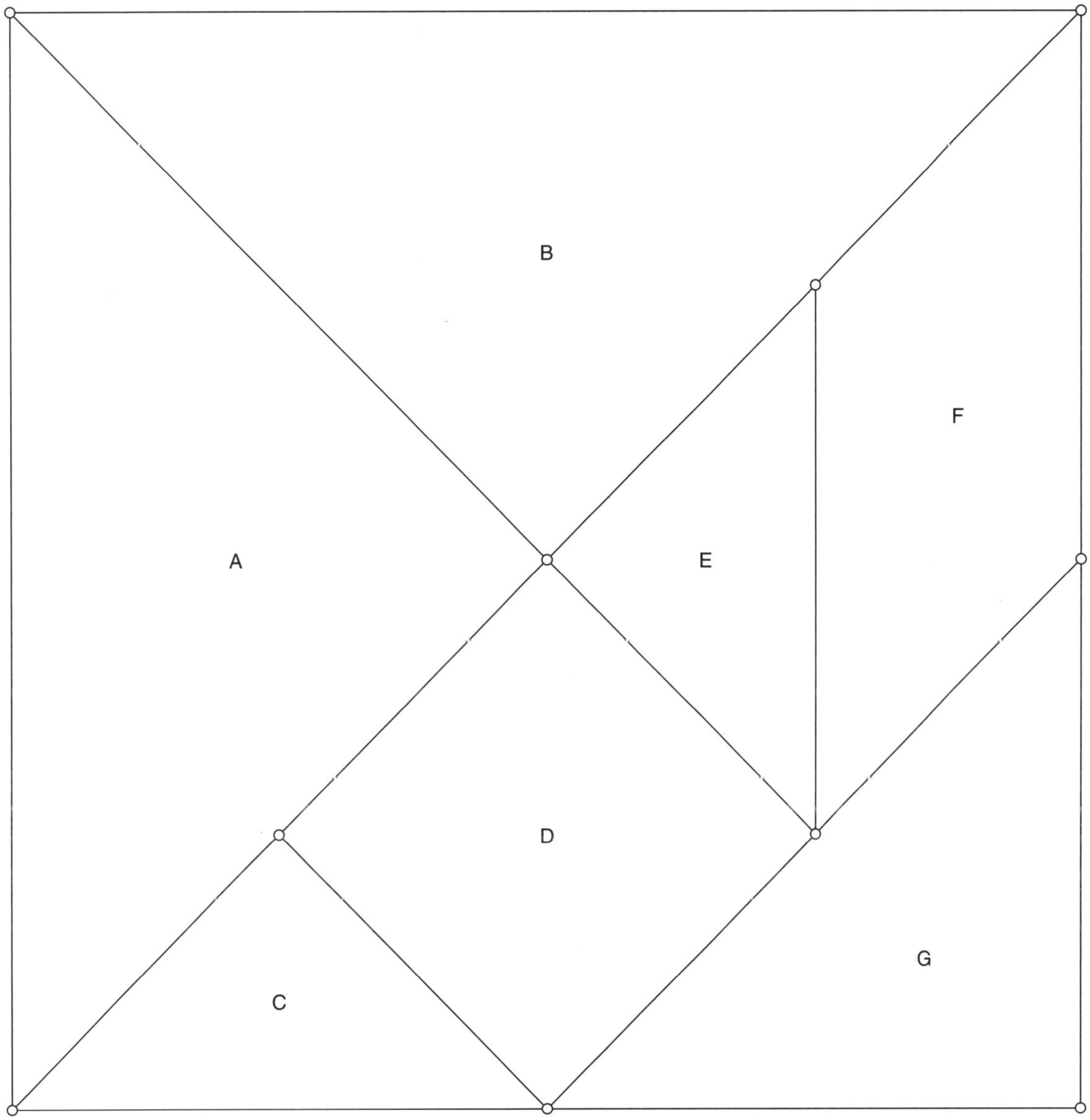

11

Tangrams

For use with Test-Day Activity 7

Name ______________________________

Use the tangram pieces you cut from Master 10 to complete exercises 1–5. Use a centimeter ruler to measure the lengths needed to answer questions 1 and 2.

1. Find the perimeter of piece A.

2. Find the area of piece D.

3. Use two pieces to create a square. Name the pieces you used.

4. Use three pieces to create a rectangle with a length of 10.6 cm and a width of 5.3 cm. Name the pieces you used.

5. Use three pieces to create a square. Name the pieces you used.

Challenge:

Use all 7 pieces to rebuild the original square. Make a sketch of your answer.

12 Constructing a Rectangular Box and Finding its Surface Area

For use with Test-Day Activity 8

Name ______________________

In this activity, you will construct a rectangular box by drawing a pattern for the box, cutting it out, folding its edges, and taping it closed.

Instructions for Constructing the Rectangular Box

a) Each of the arrangements below is a pattern for a rectangular box. Choose one of the patterns.

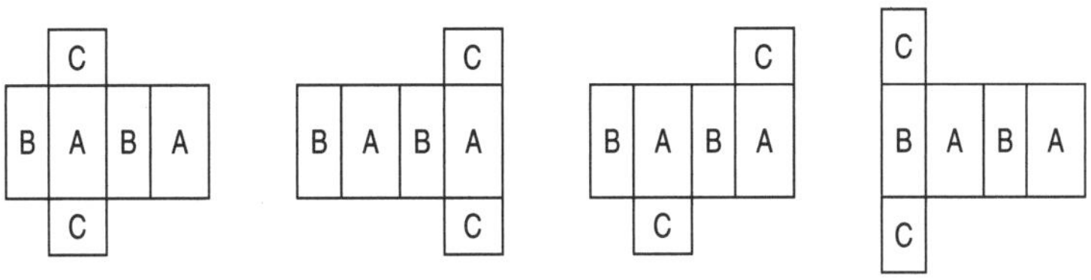

b) Cut out the dashed rectangles (A, B, and C) on Master 6-B. Use them as templates for the pattern you chose.

c) Use your template to draw your pattern on cardstock or construction paper. Be sure to mark the fold lines. Cut your pattern out. Take care to cut on the outside edge only. When you are finished cutting, you should have only one piece.

d) Use your ruler to help you find the surface area of your pattern. Write the surface area directly on the pattern. (Recall that area measurements are in square units, such as square inches.)

e) Now fold the edges and tape your pattern to make a rectangular box.

- What is the surface area of your rectangular box? How do you know?

13 Constructing a Rectangular Box and Finding its Surface Area

For use with Test-Day Activity 8

Name ______________________

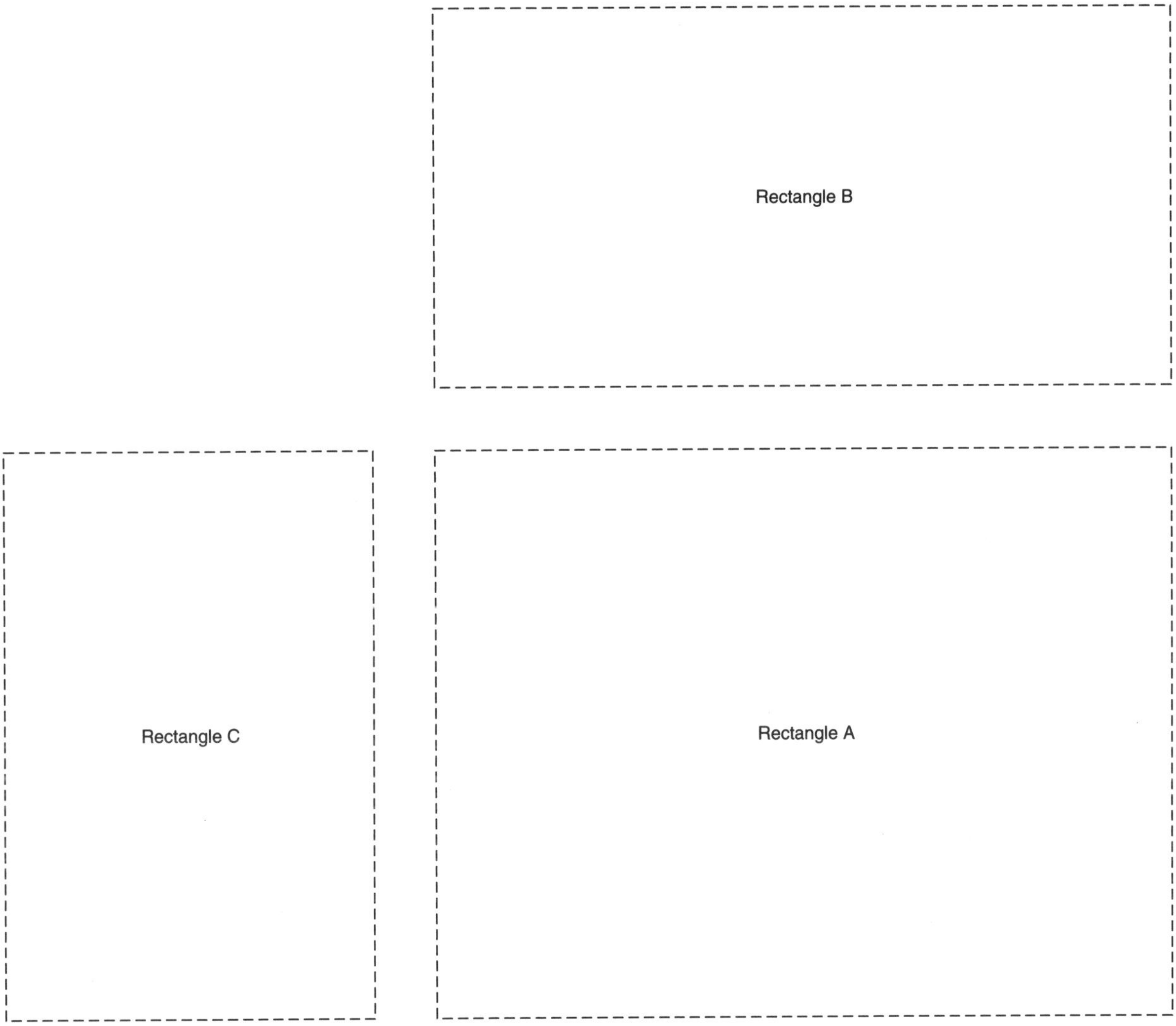

14 Creating a Math Mini-Book, Part 1

For use with Test-Day Activity 9

Name ______________________

Step 1

Fold a rectangular sheet of blank paper in half three times so that the folds make eight sections. See the figure below.

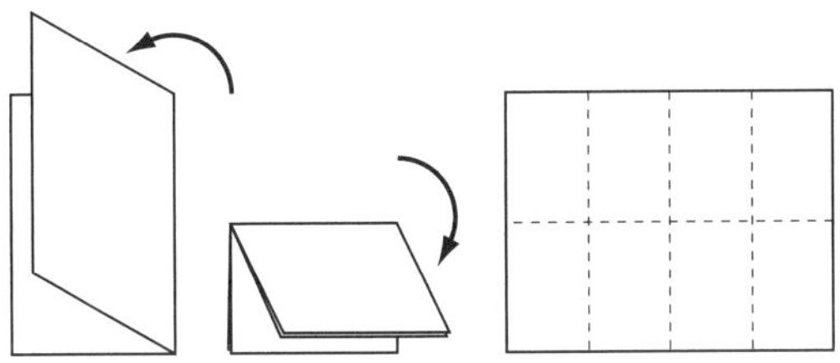

Step 2

Fold the paper in half and cut from the center of the fold as shown in the following figure.

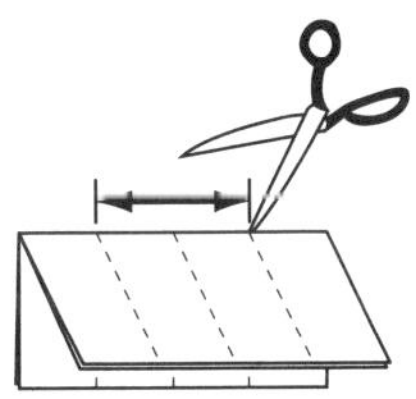

Step 3

Open the paper and fold it in half lengthwise so that the part of the paper that was cut opens as shown in the figure below.

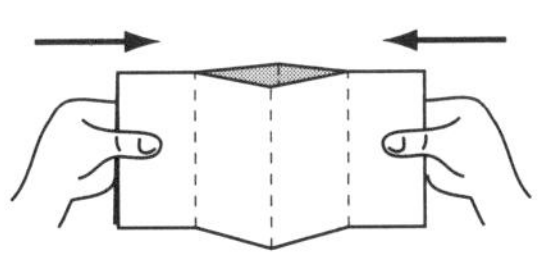

Step 4

Hold the lengthwise fold and push the ends together to make an 8-page (4-leaf) booklet.

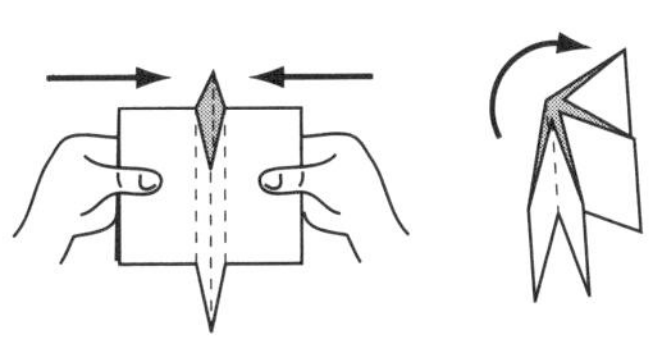

15 Creating a Math Mini-Book, Part 2

For use with Test-Day Activity 10

Name ______________________

Step 1

Choose a topic for your lesson. Use the following guidelines:

- Choose a math topic already covered in class.
- Choose a math topic that you might need to review.
- Approve your math topic with your teacher before beginning.

Write a math lesson with examples, steps, questions, and answers.

Step 2

- Write the title and author on Page 1 of the Math Mini-Book that you created in Test-Day Activity 9.
- Write the explanation of your lesson on Page 2.
- Write one or two examples on Page 3.
- Write the steps to your solutions to your examples on Pages 4, 5, and 6.
- Write questions for readers to answer on Page 7 and place the answers to your questions on the back (Page 8.)

Step 3

Share your Math Mini-Book with your friends, classmates, and people at home. Your teacher may want to post your mini-book on a bulletin board with other mini-books for students to read.

16 Measures of Central Tendency

For use with Test-Day Activity 11

Name ______________________

mean The number found when the sum of 2 or more numbers is divided by the number of addends in the sum; also called *average.*	To find the mean of the numbers 5, 6, and 10, first add. 5 + 6 + 10 = 21 Then, since there were three addends, divide the sum by three. 21 ÷ 3 = 7 The **mean** of 5, 6, and 10 is 7.
median The middle number (or the average of the two central numbers) of a list of data when the numbers are arranged in order from the least to the greatest.	1, 2, 5, 6, 7, 9, 15, 24, 36 In this list of data, 7 is the **median.**
mode The number or numbers that appear most often in a list of data.	5, 12, 32, 5, 16, 5, 7, 17 In this list of data, the number 5 is the **mode.**

17 Measures of Central Tendency

For use with Test-Day Activity 11

Name ______________________

Jeff's last eight math test scores were

85, 79, 86, 95, 86, 98, 91, and 84.

1. Find the mean, median, and mode of his scores.

2. Which measure of central tendency (mean, median, or mode) best describes Jeff's test performance in math? Explain your reasoning.

TEST-DAY ACTIVITY MASTER

18 Measures of Central Tendency

For use with Test-Day Activity 11

Name ______________________

1. In a particular town, there are a large number of medium-priced homes and a few very high-priced homes. In a recent month, the following sale prices of homes in the town were recorded.

$210,000

$215,000

$195,000

$925,000

$225,000

For a potential buyer of a home in the town, which measure of central tendency provides more useful information about the price of a typical home, the mean or the median of these prices? Explain your answer.

2. Before the new bowling alley opened, the equipment purchaser gathered information about local bowlers. Calculating the results, the equipment purchaser found that the median weight of bowling balls used was 12 pounds, the mean weight was 12.6 pounds, and the two most commonly used weights (the modes) were 10 and 16 pounds. To guide the equipment purchaser in his ordering decisions, which measure provides the most useful information, the median, mean, or mode? Explain your answer.

19 Geometry in the Real World, Part 1

For use with Test-Day Activity 12

Name ____________________

Shape	Name	Examples in the Real World
	**Square**	
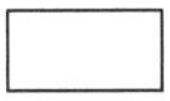	**Rectangle**	
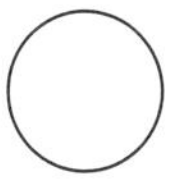	**Circle**	
	**Triangle**	
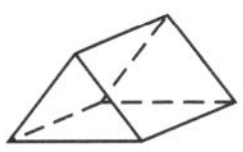	**Triangular Prism**	
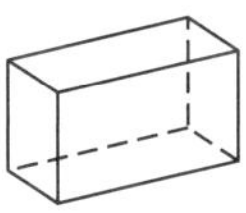	**Rectangular Prism**	
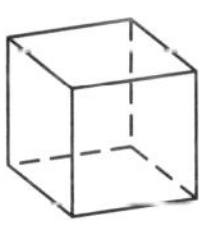	**Cube**	
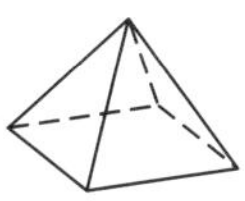	**Pyramid**	
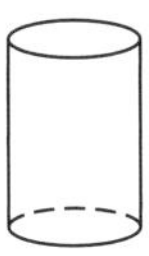	**Cylinder**	
	**Cone**	
	**Sphere**	

20 Geometry in the Real World, Part 2

For use with Test-Day Activity 13

Name ______________________________

Geometry in the Real World

Poster Project

Due Date: __________

Materials needed:
Poster board; glue; scissors; and magazines, books, photographs, or the Internet

Instructions:
Use magazines, photographs, or the Internet to find pictures of where geometry is found in nature, art, and architecture. Use the list you created on Master 19 and the list your class created for ideas. Find one picture for each of the shapes listed on the master. Paste the pictures neatly onto the poster board. Next to each picture write the name of the shape that is in the picture. Make sure that each picture shows geometry being used in nature, art, or architecture. Be creative with your poster.

21 Selecting Appropriate Methods and Tools, Part 1

Name ______________________

For use with Test-Day Activity 14

Choose an appropriate method: estimation or exact. Choose an appropriate tool: mental math, calculator, or paper/pencil.

1. The cost of making a long distance phone call is \$2.00 for the first 20 minutes plus \$0.07 for each additional minute. What would be the cost of a 45-minute phone call?

2. Judy is having a dinner party for 6 people. She has decided to use a recipe that only feeds 3 people, so she will need to double the recipe. Help Judy by doubling the recipe below. Rewrite the doubled amounts in the space provided.

$1\frac{3}{4}$ cups flour ____

$\frac{1}{4}$ teaspoon soda ____

2 teaspoons baking powder ____

$\frac{1}{2}$ teaspoon salt ____

$\frac{1}{3}$ cup shortening ____

$\frac{2}{3}$ cup sugar ____

2 eggs ____

1 cup sliced bananas ____

3. You are riding in a car that is traveling at a rate of 60 miles per hour. About how far will you travel in 15 minutes?

22 Two-Dimensional Representations of Three-Dimensional Objects

For use with Test-Day Activity 15

Name ______________________

Three rectangular shoe boxes were used to build this shape.

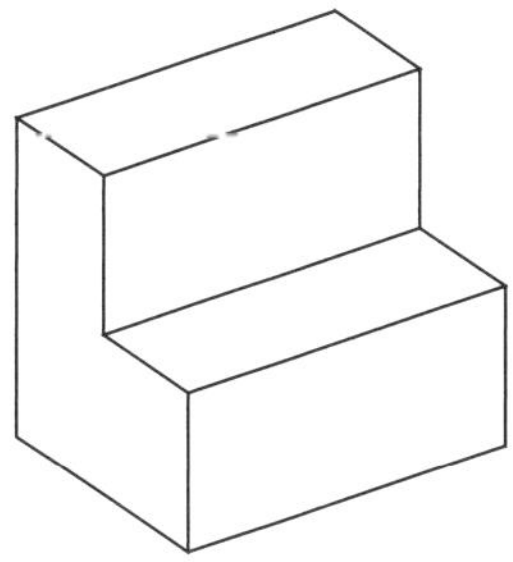

Below are sketches of this shape when viewed from the front, top, and right side.

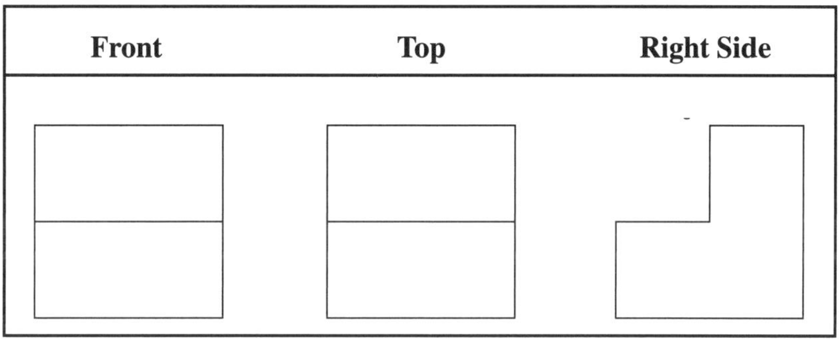

Four rectangular shoe boxes were used to build this shape.

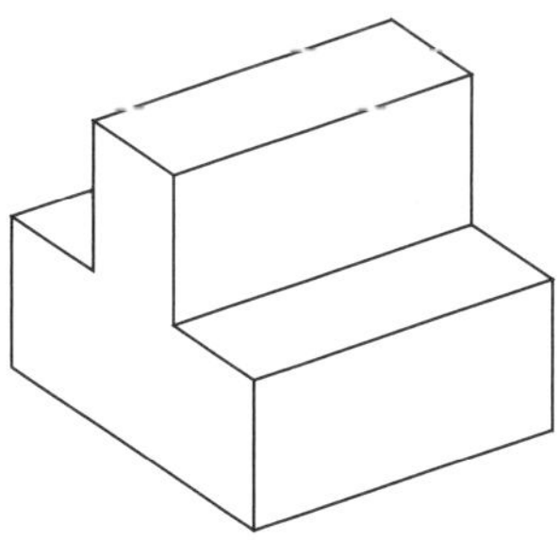

Sketch the front, top, and right side views of the shape.

23

Two-Dimensional Representations of Three-Dimensional Objects

For use with Test-Day Activity 15

Name ____________________

Five cubes were used to build this shape.

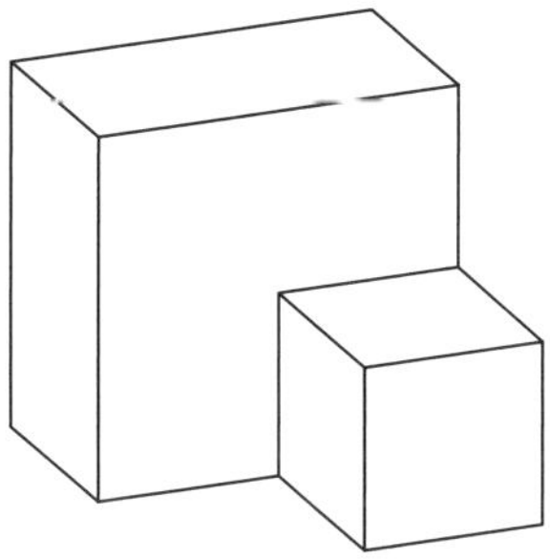

Below are sketches of the shape when viewed from the front, top, and right side.

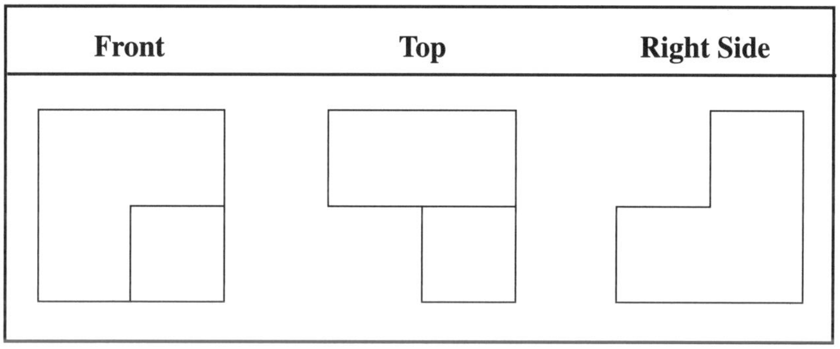

Seven cubes were used to build this shape.

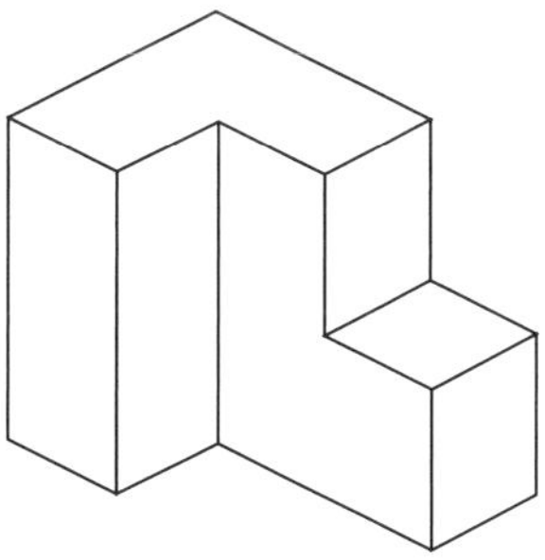

Sketch the front, top, and right side views of the shape.

24 Estimating and Measuring Capacity

For use with Test-Day Activity 16

Name ____________________

Activity Instructions

1. Look at the 1-pint container. Estimate how many cups of water it will take to fill that container. Record your estimate in the chart on Master 25.

2. Measure one cup of water and pour it into the 1-pint container (use the funnel if needed). Repeat this process until the container is full. Record the actual number of cups in a pint in the chart on Master 25.

3. Look at the 1-quart container. Estimate how many cups of water it will take to fill that container. Record your estimate in the chart on Master 25.

4. Measure one cup of water and pour it into the 1-quart container (use the funnel if needed). Repeat this process until the container is full. Record the actual number of cups in a quart in the chart on Master 25.

5. Look at the 1-gallon container. Estimate how many cups of water it will take to fill that container. Record your estimate in the chart on Master 25.

6. Measure one cup of water and pour it into the 1-gallon container (use the funnel if needed). Repeat this process until the container is full. Record the actual number of cups in a gallon in the chart on Master 25.

7. Use the information you have gathered to calculate the number of pints in a quart, the number of pints in a gallon, and the number of quarts in a gallon. Record your answers in the chart on Master 25.

25 Estimating and Measuring Capacity

For use with Test-Day Activity 16

Name ______________________

	cups	pints	quarts
1 pint =	Estimate ____ Actual ____		
1 quart =	Estimate ____ Actual ____	Actual ____	
1 gallon =	Estimate ____ Actual ____	Actual ____	Actual ____

	Estimate	Actual
Mystery container	**#1** ____	____
	#2 ____	____
	#3 ____	____

TEST-DAY ACTIVITY MASTER

26

Using a Calculator to Explore Exponents

For use with Test-Day Activity 17

Name ______________________

Using a Calculator

Calculators can be used to evaluate exponential expressions. Use the following steps to learn how this can be done:

a) Enter a number.

b) Press [×] [=]

c) Keep pressing [=] until the desired power has been reached

Example: 3^5 = [3] [×] [=] [=] [=] [=]

3^1 3^2 3^3 3^4 3^5

- Try 4^6 on your calculator. Make sure you get 4096.

Notice that the number of times you press [=] is one less than the exponent.

27 Using a Calculator to Explore Exponents

For use with Test-Day Activity 17

Name ______________________

Use a calculator to evaluate the following exponential expressions. Write the calculator keystrokes in the space provided.

1. 5^5 = ________

keystrokes________

2. 11^3 = ________

keystrokes________

3. 2^{10} = ________

keystrokes________

4. 8^4 = ________

keystrokes________

Evaluate the exponential expression.

5. 3^6 = ____

6. 9^7 = ____

7. 4^9 = ____

8. 10^2 = ____

28 The Sum of the Angles of a Triangle

For use with Test-Day Activity 18

Name ______________________________

The Sum of the Angles of a Triangle

1. Use a pencil and a ruler to draw two different triangles on one sheet of paper. Each side of both triangles should be longer than one inch. The triangles may be obtuse, right, or acute.

2. Use a compass to draw an arc across each angle of the first triangle. Center the compass on the vertex of each angle and do not change the radius. Make sure that the radius is small enough that the arcs do not intersect (cross).

3. Use scissors to cut out the first triangle. Then cut off each angle along the arc, and fit the three cut angles together to make a half circle.

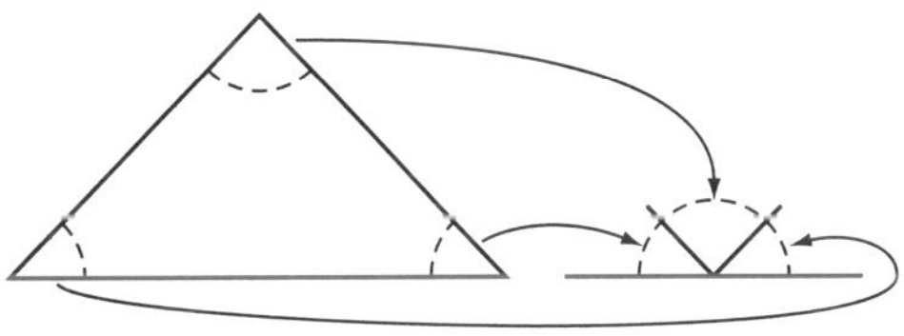

4. Repeat steps 2 and 3 with your second triangle.

Since the "corners" of a triangle fit together to form a half circle, the sum of the measures of the angles of a triangle is 180°. This is true for any triangle.

Try This Find the measure of the third angle.

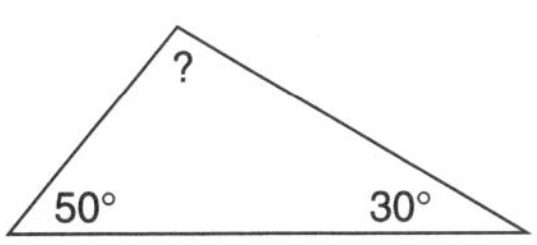

29

The Sum of the Angles of a Quadrilateral

For use with Test-Day Activity 18

Name ______________________

The Sum of the Angles of a Quadrilateral

1. Use a pencil and a ruler to draw two different quadrilaterals on one sheet of paper. Each side of both quadrilaterals should be longer than one inch.

2. Use a compass to draw an arc across each angle of the first quadrilateral. The arcs should have the same radius and should not intersect.

3. Use scissors to cut out the first quadrilateral. Then cut off each angle along the arc and fit the four cut angles together to make a circle.

4. Repeat steps 2 and 3 with your second quadrilateral.

> Since the "corners" of a quadrilateral fit together to form a circle, the sum of the measures of the angles of a quadrilateral is 360°. This is true for any quadrilateral.

Try This Find the measure of the fourth angle.

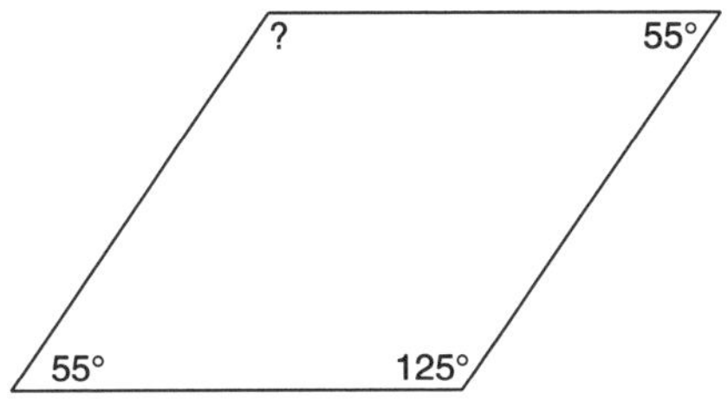

TEST-DAY ACTIVITY MASTER

30 The Sum of the Angles of a Triangle and a Quadrilateral

For use with Test-Day Activity 18

Name ______________________

Find each missing angle.

1.
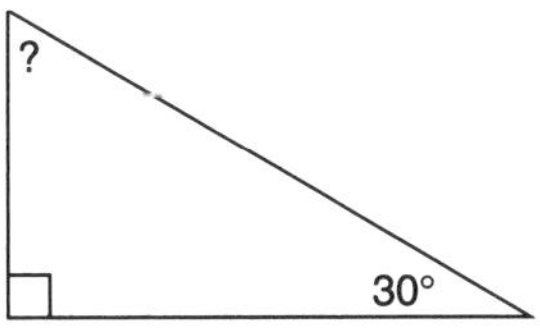

2.
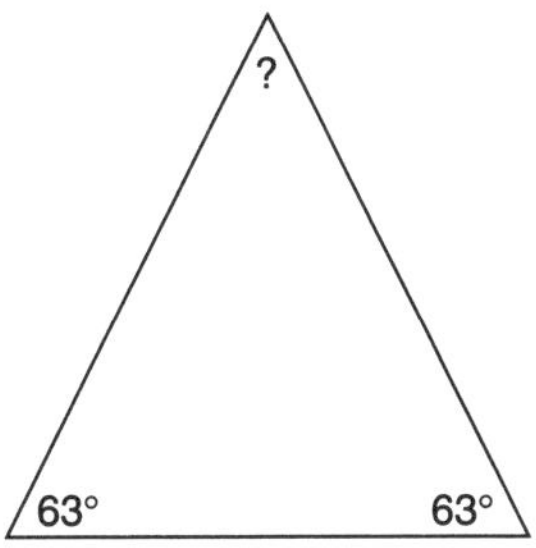

3.
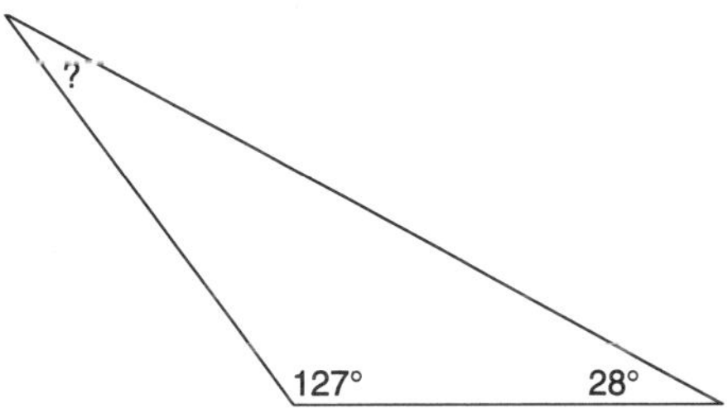

4.
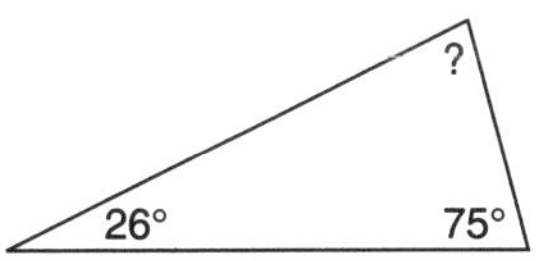

5.
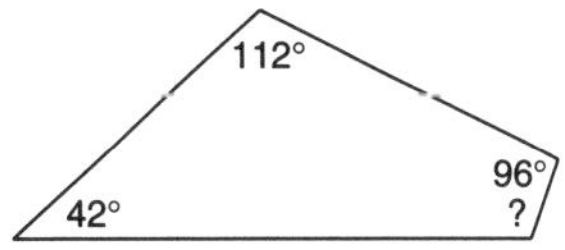

6.
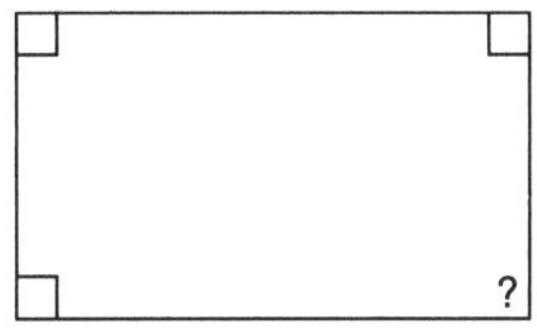

7.
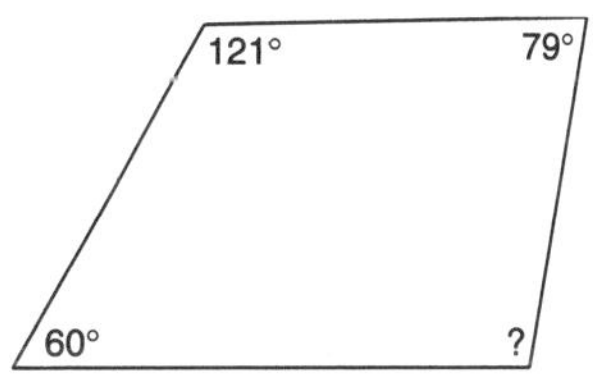

8.
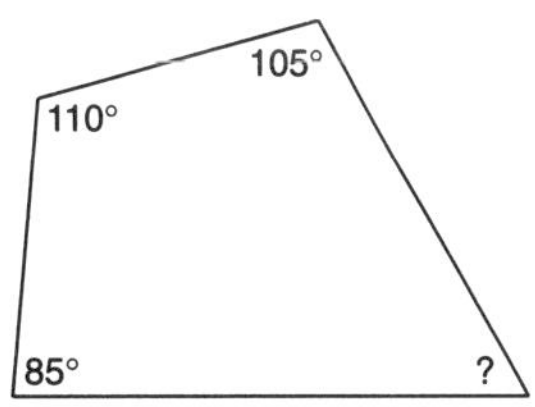

31

Perimeter, Area, and Volume

For use with Test-Day Activity 19

Name ______________________

Perimeter of a Rectangle		
Length (in.)	**Width (in.)**	**Perimeter (in.)**
2	1	6
3	2	10
4	3	14
5	4	18
6	5	22
l	w	P

1. Find the perimeter of a rectangle with a length of 7 in. and a width of 6 in.

2. Assume the length of a rectangle is l and its width is w. Use the pattern in the table to complete the formula for the perimeter of a rectangle: $P =$ ____.

Area of a Rectangle		
Length (cm)	**Width (cm)**	**Area (cm^2)**
2	1	2
3	2	6
4	3	12
5	4	20
6	5	30
l	w	A

3. Find the area of a rectangle with a length of 7 cm and a width of 6 cm.

4. Assume the length of a rectangle is l and its width is w. Use the pattern in the table to complete the formula for the area of a rectangle: $A =$ ____.

TEST-DAY ACTIVITY MASTER

32

Perimeter, Area, and Volume

For use with Test-Day Activity 19

Name ______________________

Area of a Square	
Side (m)	**Area (m^2)**
1	1
2	4
3	9
4	16
5	25
s	A

1. Find the area of a square whose side is 6 m long.

2. Use the pattern in the table to complete the formula for the area of a square whose side is s: A = ____.

Volume of a Rectangular Prism			
Length (ft)	**Width (ft)**	**Height (ft)**	**Volume (ft^3)**
2	1	3	6
3	2	4	24
4	3	5	60
5	4	6	120
6	5	7	210
l	w	h	V

3. Find the volume of a rectangular prism with a length of 7 ft, a width of 6 ft, and a height of 8 ft.

4. Assume the length of a rectangular prism is l, its width is w, and its height is h. Use the pattern in the table to complete the formula for the volume of a rectangular prism: V = ____.

33 Selecting Appropriate Methods and Tools, Part 2

Name ______________________

For use with Test-Day Activity 20

For each problem below, write three responses. First, choose the appropriate form of the answer: estimated or exact. Second, choose the appropriate method: mental calculation, paper and pencil calculation, or calculator usage. Third, write the answer to the question.

1. Susan needs to make 5 tablecloths for an upcoming school banquet. Each tablecloth requires $3\frac{1}{3}$ yards of fabric. How many yards of fabric does she need to buy?

2. Stuart is making chocolate chip cookies for a bake sale. He already has $4\frac{1}{2}$ cups of chocolate chips in his kitchen. He is planning to bake 3 batches of cookies that require $2\frac{1}{4}$ cups of chocolate chips each. Does he have enough chocolate chips or does he need to buy more?

3. Laura walked on her treadmill for $\frac{3}{4}$ hour on Monday, $\frac{1}{2}$ hour on Tuesday, and $\frac{2}{3}$ hour on Wednesday. About how long did she exercise over the 3 days?

TEST-DAY ACTIVITY MASTER

34 Using Positive and Negative Integers in the Real World

For use with Test-Day Activity 21

Name ______________________

In the game of golf, *par* is the score an expert golfer is expected to make on a given hole. The object of the game is to get the ball in the hole, making the fewest strokes possible. One way to keep score in the game is to keep track of the number of strokes above or below par a player makes on each hole.

On the scorecard below, Jim scored +1 (or one stroke above par) on Hole 1. This means that it took Jim 6 strokes to get the ball in the hole.

Hole	1	2	3	4	5	6	7	8	9
Par	5	4	3	4	4	3	4	5	4
Jim's score	+1	−1	0	+2	0	+1	−1	−2	−1

1. On what hole did Jim make the most strokes under par?

2. On what hole did Jim make the most strokes over par?

3. How many strokes did it take Jim to get the ball in the hole on Hole 7?

4. Was Jim's score for the entire 9 holes above or below par?

35

Using Positive and Negative Integers in the Real World

For use with Test-Day Activity 21

Name ______________________________

Sam records deposits into and withdrawals from his savings account. He uses a negative number to show a withdrawal:

March 29	$314
April 7	–$129
April 14	–$67
May 1	$261
May 17	–$296

1. On what date did Sam make the largest deposit?

2. On what date did Sam make the largest withdrawal?

3. Has Sam deposited more than he has withdrawn on the five dates listed? Explain your answer.

36

Mass

For use with Test-Day Activity 22

Name ______________________

Mass

1 kilogram (km) = 1000 grams (g)
1000 milligrams (mg) = 1 gram (g)

Benchmarks

The mass of a grain of table salt is about one milligram (mg).	The mass of a dollar bill is about one gram (g).	The mass of a baseball bat is about one kilogram (kg).
1 milligram 	 1 gram	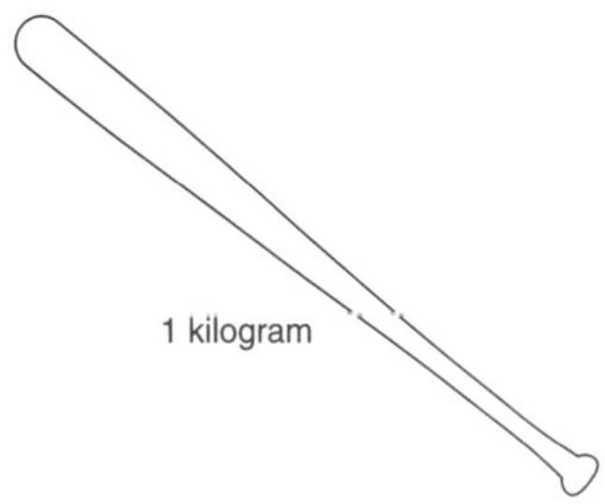 1 kilogram

37

Mass

For use with Test-Day Activity 22

Name ______________________

Choose the reasonable unit of measure for each. Write mg, g, or kg.

1. The mass of a bread crumb is about 1 ____.

2. The mass of your math book is about 1 ____.

3. The mass of a dozen large apples is about 3 ____.

4. The mass of 4 potato chips is about 2 ____.

Choose the unit you would use to measure the mass of each object. Write mg, g or kg.

5. a jellybean ____

6. a truck ____

7. a coffee table ____

8. a piece of paper ____

9. a grain of rice ____

10. a tennis ball ____

Activity Masters

The following Activity Masters are referenced in the *Saxon Math 7/6* textbook. The Activity Masters should be copied for students to use during the appropriate lesson or investigation. We suggest that the fraction manipulatives masters (Activity Masters 3–7) be copied on five different colors of paper or that students color-code their copies of the fraction manipulatives before cutting them. Some of these masters are worksheets; their answers can be found in the back of this book.

ACTIVITY MASTER

1 Test-Scores Line Graph

For use with Lesson 18

Name ______________________

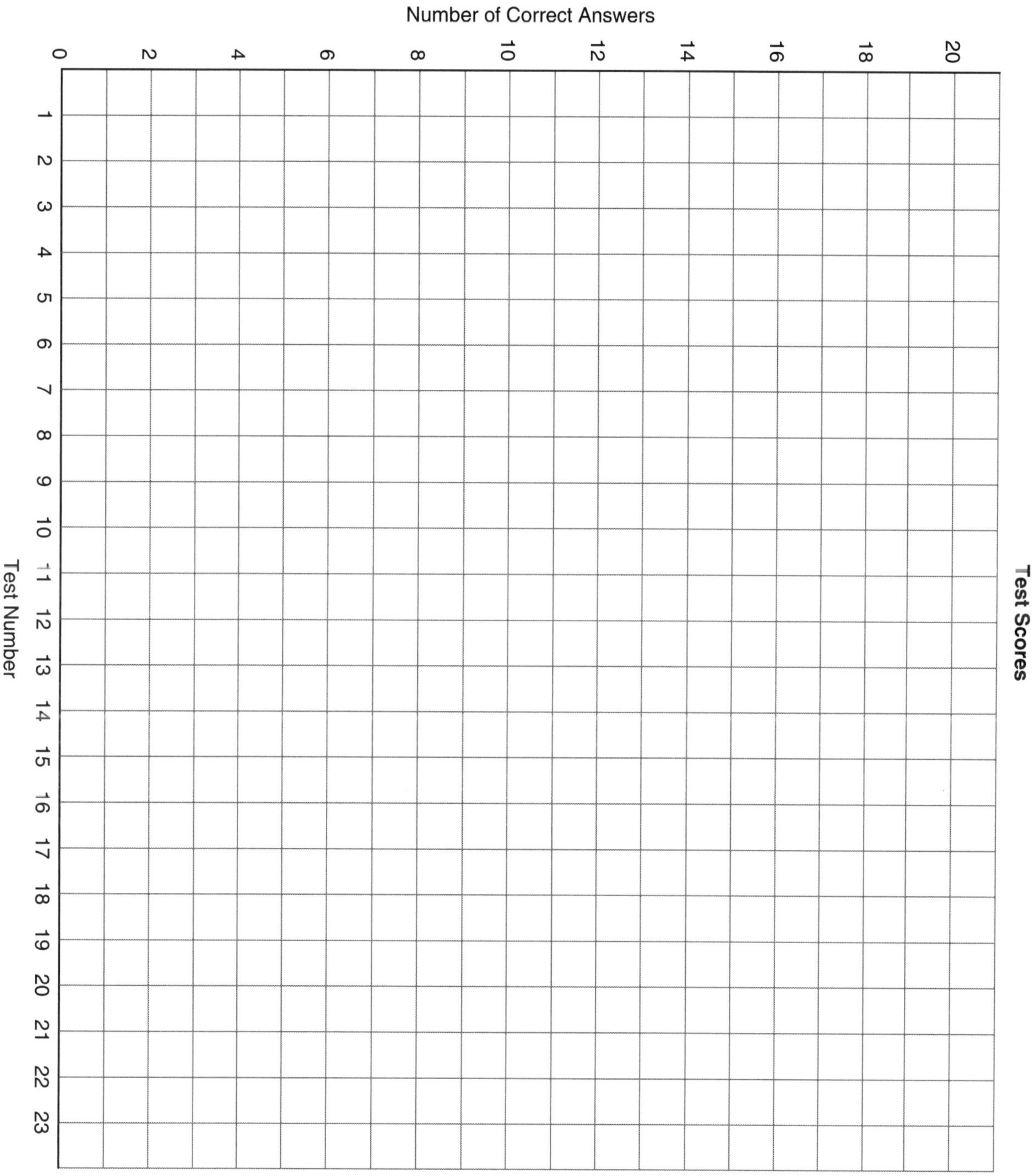

2

Hundred Number Chart

For use with Lesson 19

Name ____________________

1	2	3	4	5	6	7	8	9	10
11	12	13	14	15	16	17	18	19	20
21	22	23	24	25	26	27	28	29	30
31	32	33	34	35	36	37	38	39	40
41	42	43	44	45	46	47	48	49	50
51	52	53	54	55	56	57	58	59	60
61	62	63	64	65	66	67	68	69	70
71	72	73	74	75	76	77	78	79	80
81	82	83	84	85	86	87	88	89	90
91	92	93	94	95	96	97	98	99	100

3

Halves

For use with Investigation 2

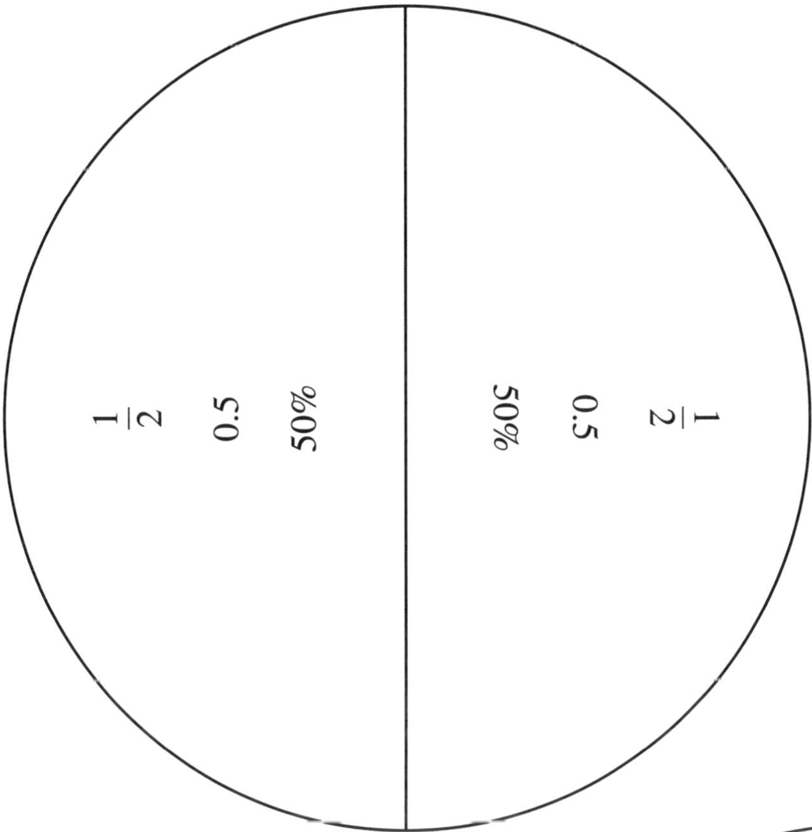

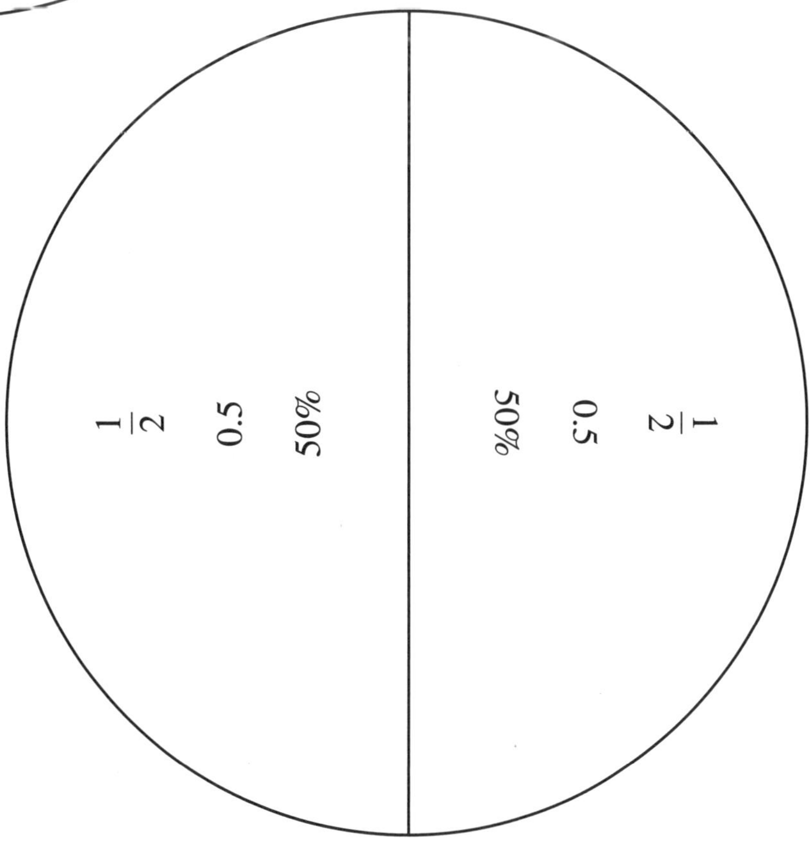

ACTIVITY MASTER

4

Thirds

For use with Investigation 2

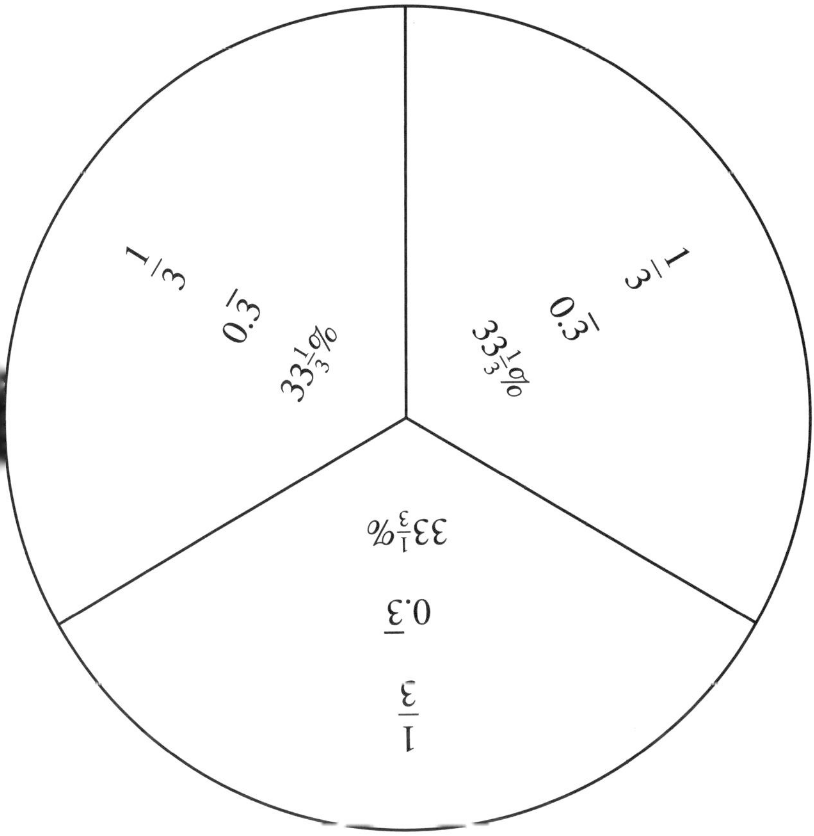

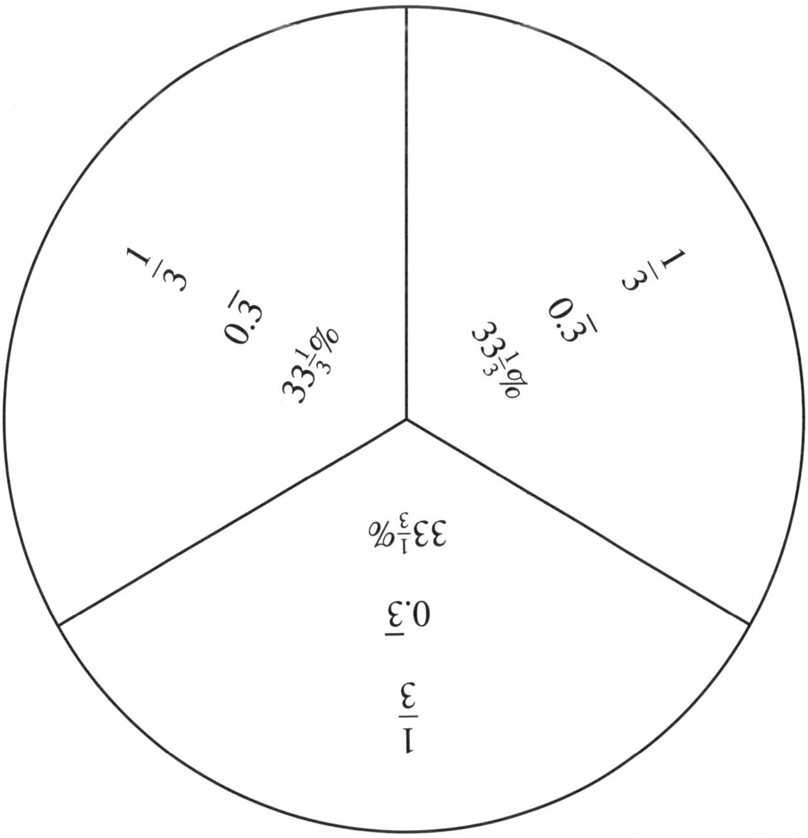

5 Fourths

For use with Investigation 2

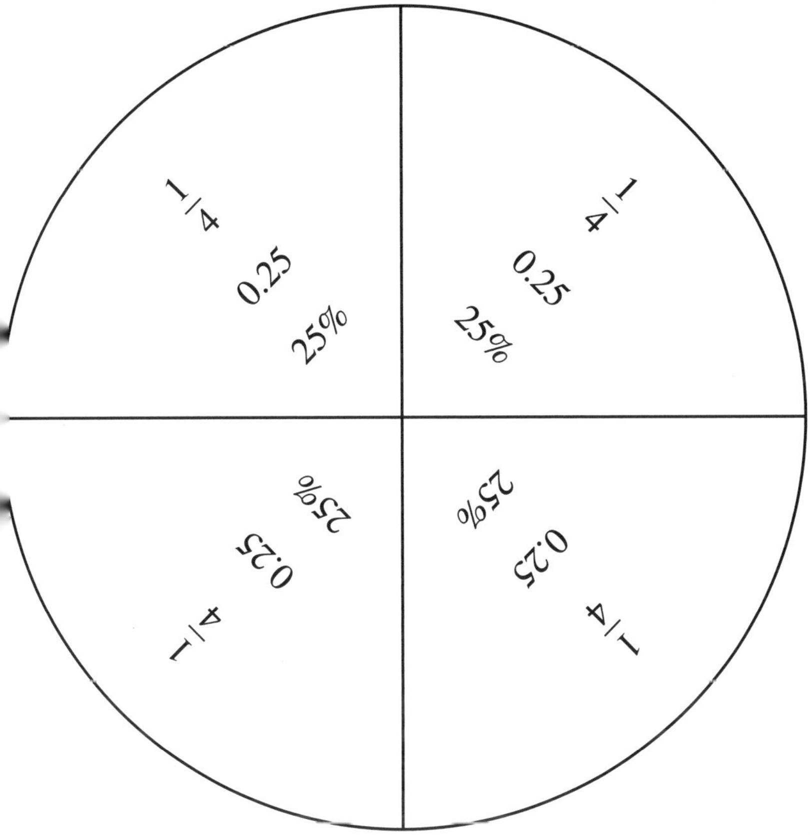

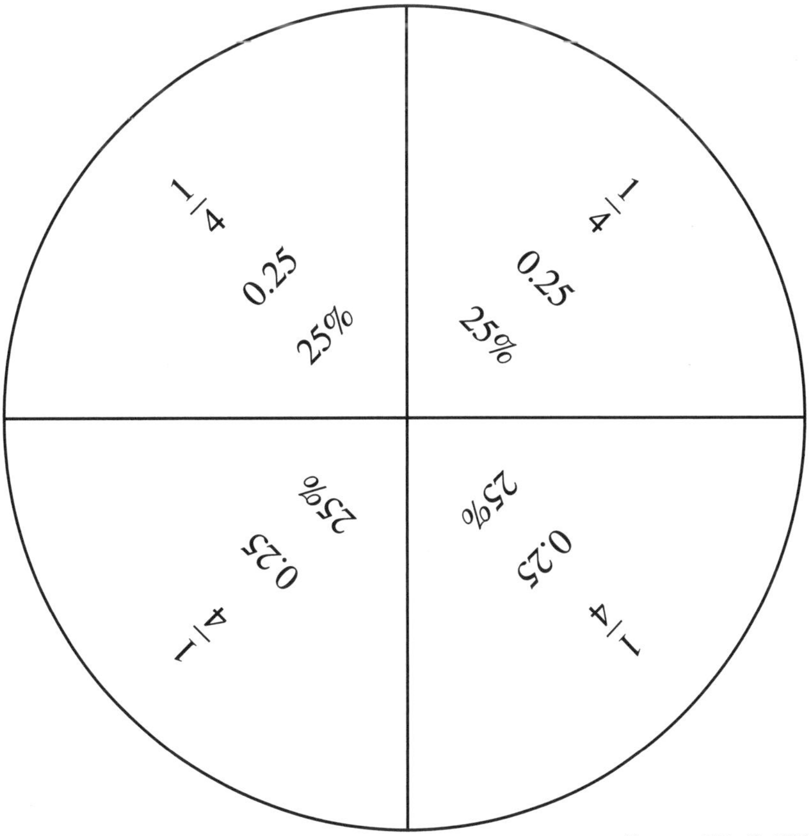

6

Sixths

For use with Investigation 2

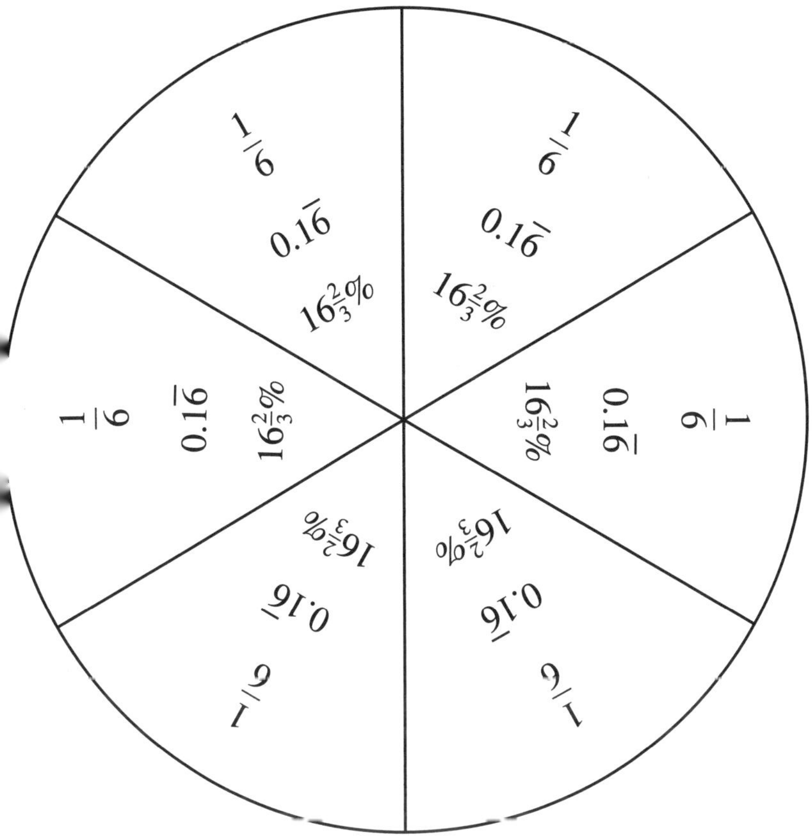

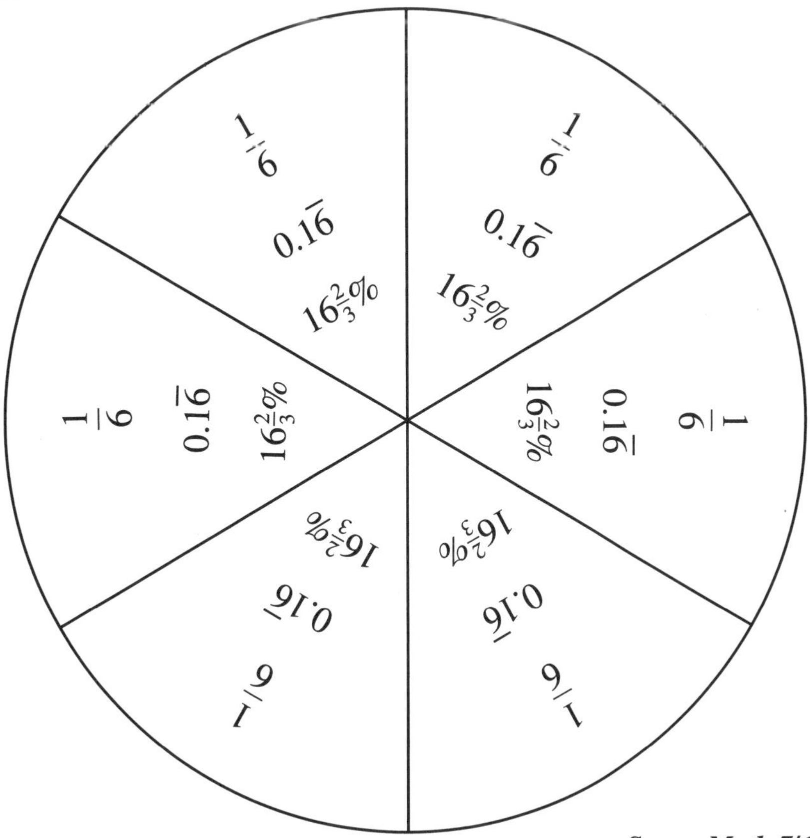

ACTIVITY MASTER

7

Eighths

For use with Investigation 2

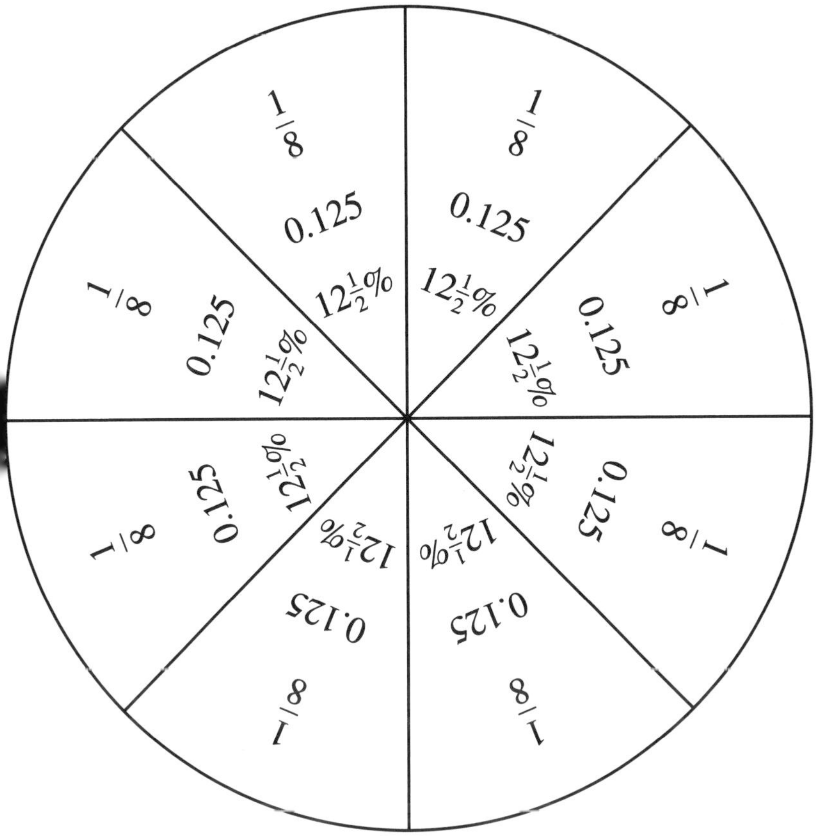

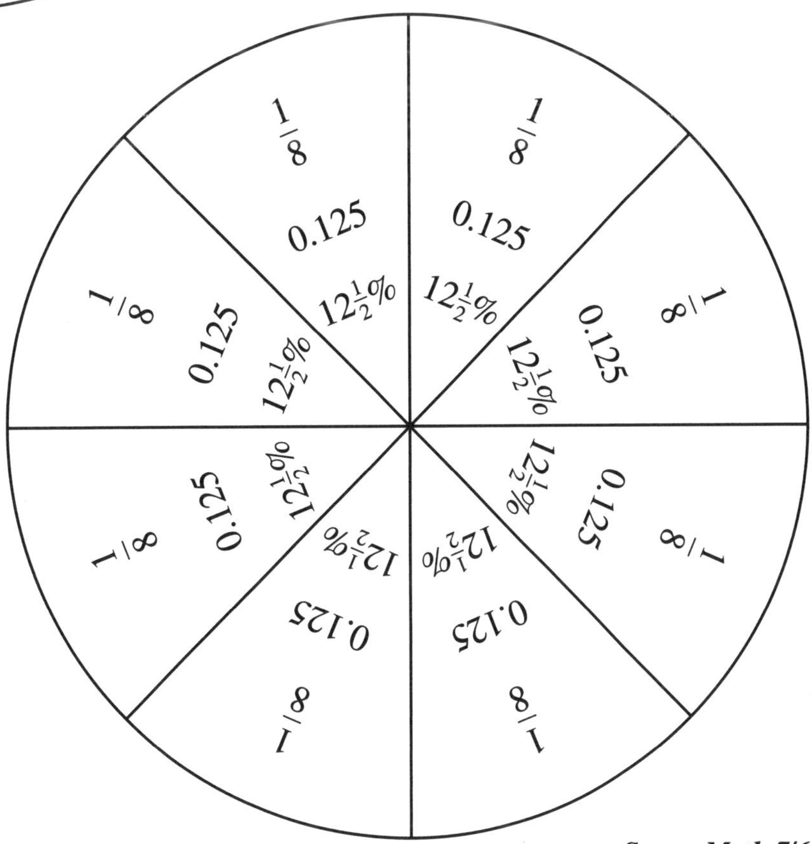

ACTIVITY MASTER

8 Measuring Angles

For use with Investigation 3

Name ______________________

se a protractor to find the following angle measures:

1. m∠*AOB* _____ **2.** m∠*AOC* _____ **3.** m∠*EOD* _____

4. m∠*AOD* _____ **5.** m∠*EOB* _____ **6.** m∠*DOG* _____

7. m∠*EOF* _____ **8.** m∠*EOG* _____ **9.** m∠*DOF* _____

10. m∠*EOA* _____ **11.** m∠*DOB* _____ **12.** m∠*COF* _____

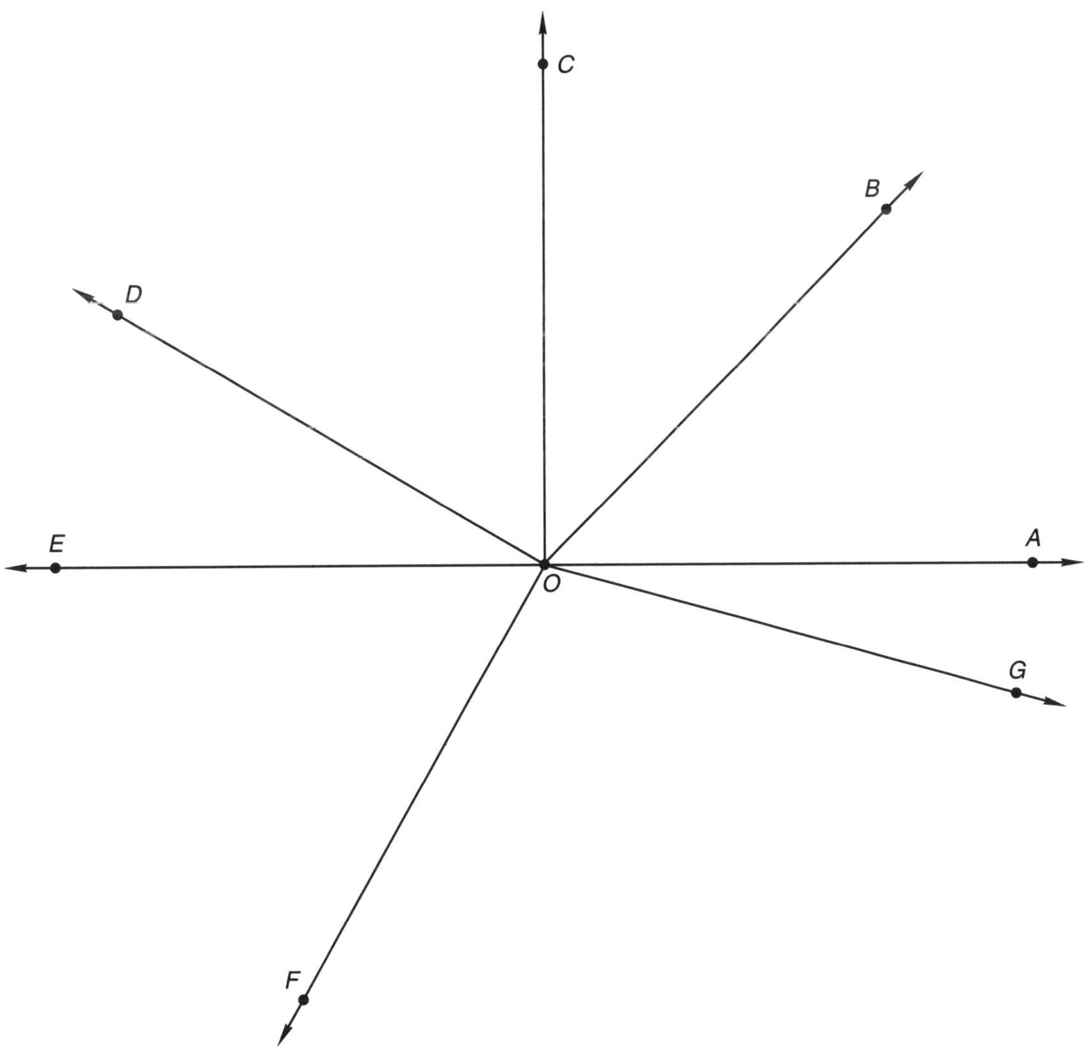

9 Finding the Number of Diameters in a Circumference

For use with Lesson 47

Name ______________________________

Part 1: Estimates

Object	Approximate Number of Diameters in the Circumference

Part 2: Measures

Object	Circumference	Diameter	$\frac{\text{Circumference}}{\text{Diameter}}$

10

Coordinate Plane

For use with Investigation 7

Name ______________________________

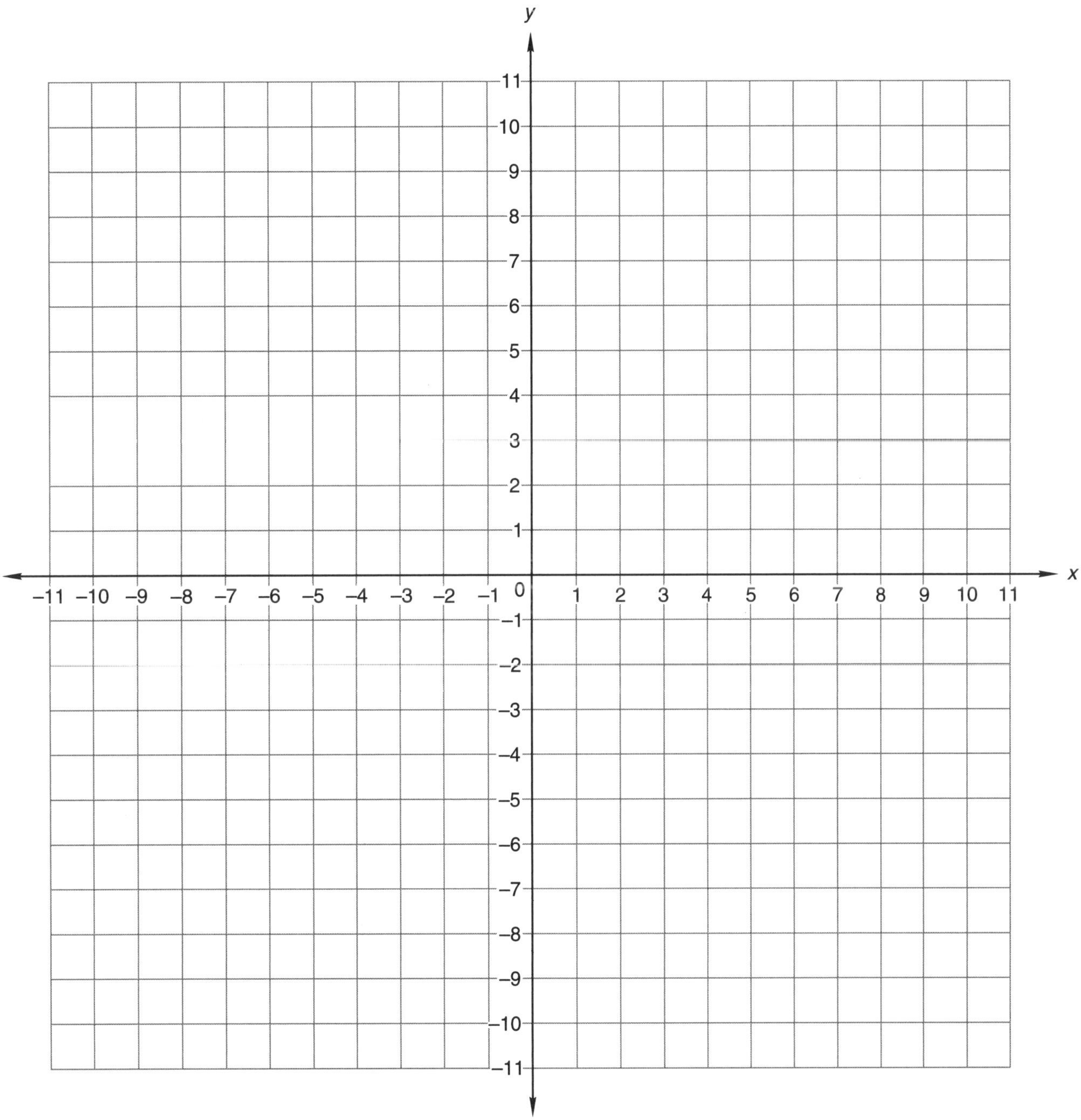

11 Constructing Bisectors

For use with Investigation 8

Name ______________________

Section A

Section B

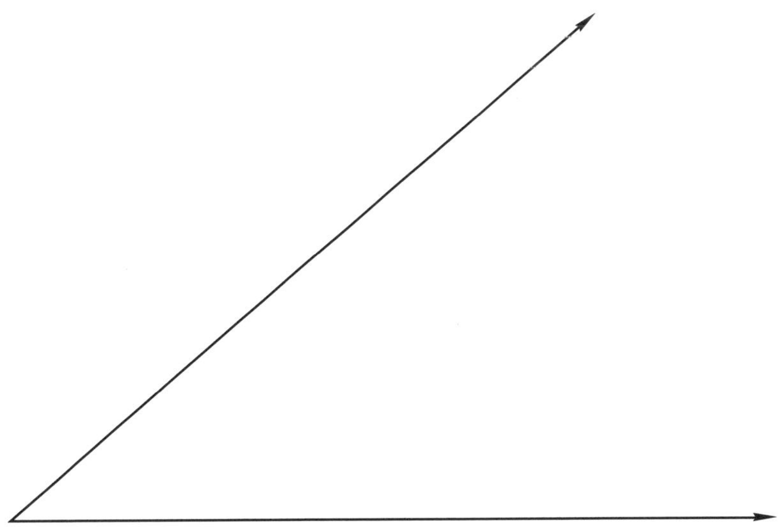

12 Scale Model

For use with Investigation 11

Art by Dan Shippey of Delta 7 Studios

Instructions

Step 1: Carefully cut out each piece on both pages. Label the back of each piece with its corresponding letter label.

Step 2: Glue (B), (D), (E), and (G) in a loop as instructed on each piece. Carefully cut (F) along the dashed line and glue as indicated.

Step 3: Attach (A) to (B) by folding the tabs on (A) down and gluing them to (B). Attach (C) to (D) and (H) to (G) in the same manner.

Step 4: Attach (G) to (F) by folding out the black tabs on (G) and gluing them to (F). Attach (B) to (C) in the same manner. The tabs on (B) should line up with the white rectangles on (C).

Step 5: Locate the tab labeled "X" on (D). Glue this tab to the corresponding rectangle labeled "X" on (E). Then glue the rest of the tabs on (D) to (E). Attach (F) to (E) in the same manner.

Assembly Diagram

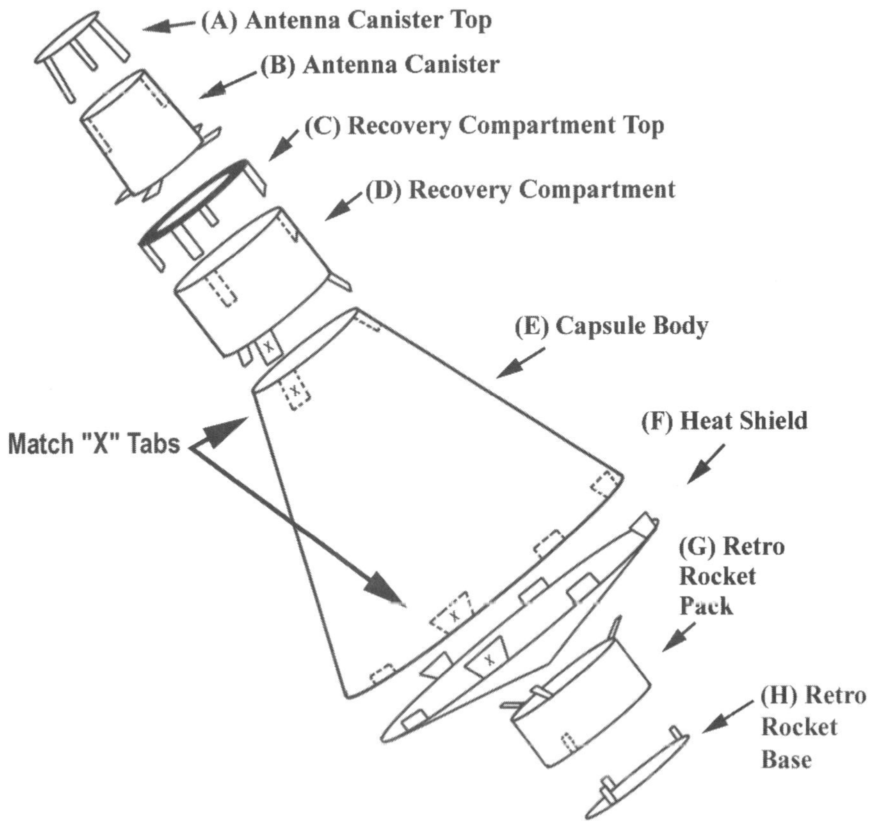

(A) Antenna Canister Top

(B) Antenna Canister

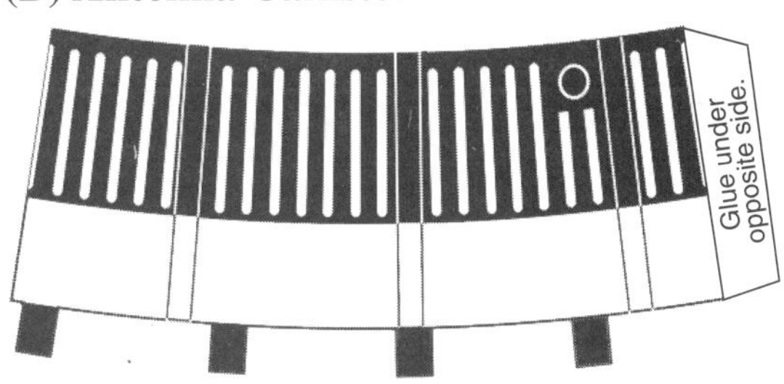

(C) Recovery Compartment Top

(D) Recovery Compartment

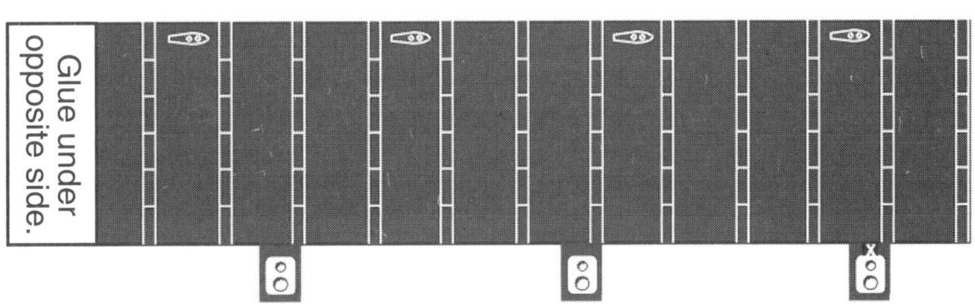

13

Scale Model

For use with Investigation 11

Art by Dan Shippey of Delta 7 Studios

(E) Capsule Body

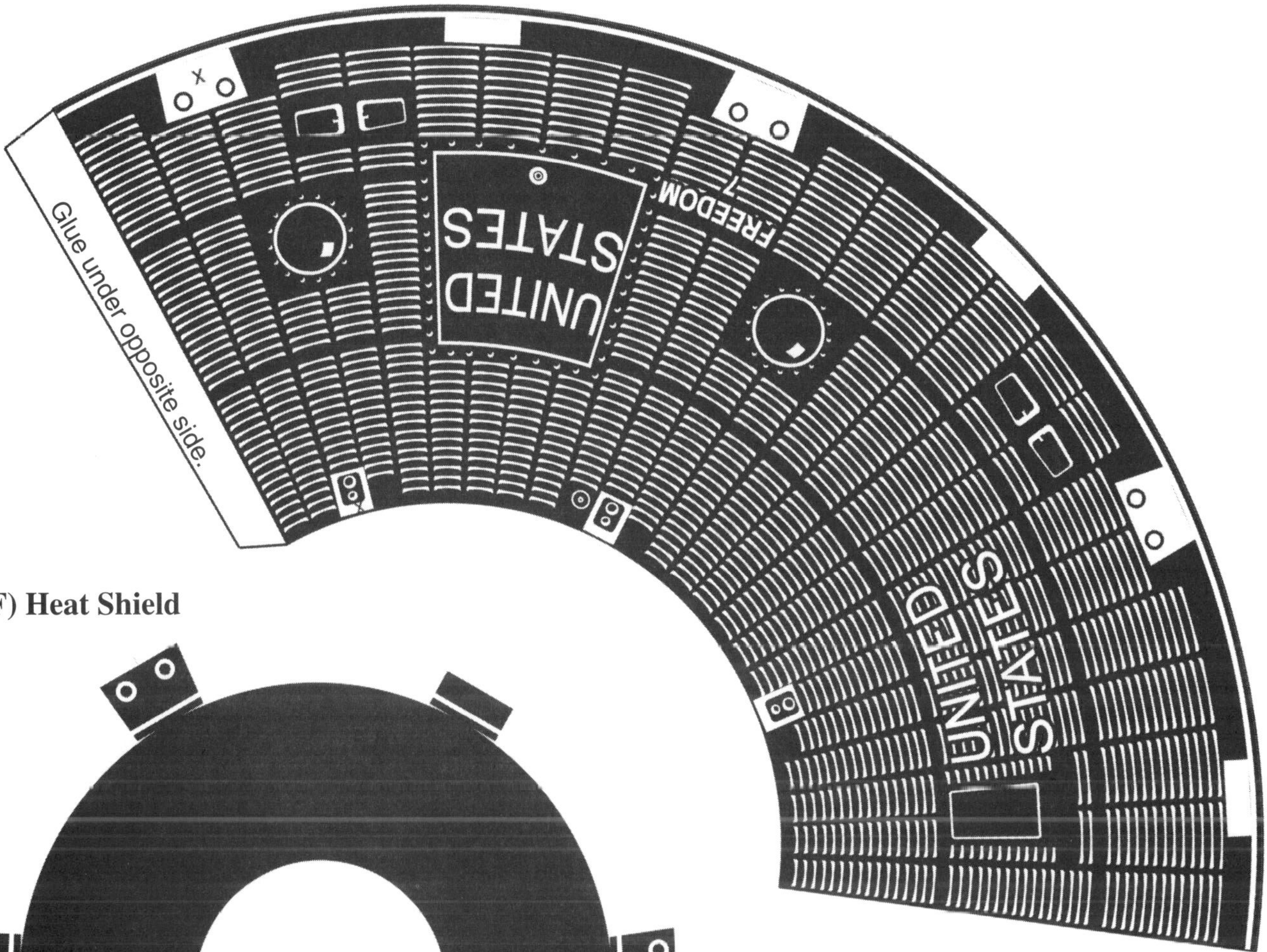

(F) Heat Shield

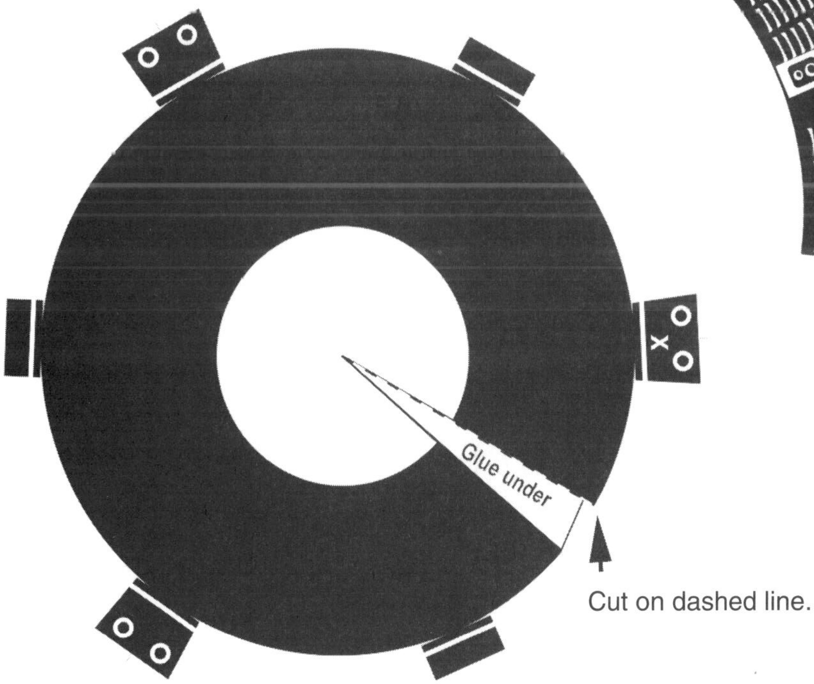

(H) Retro Rocket Base

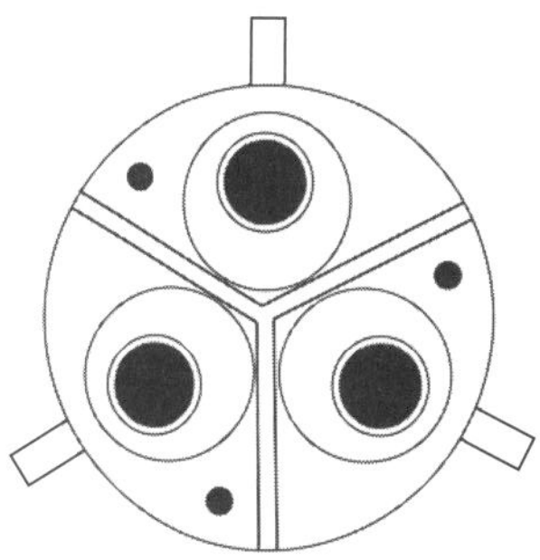

(G) Retro Rocket Pack

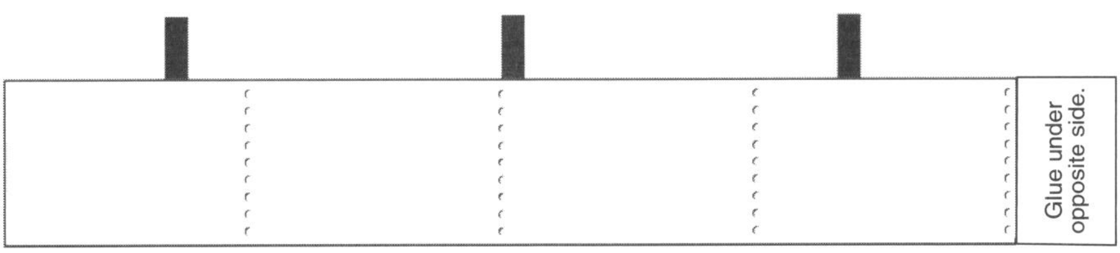

ACTIVITY MASTER

14 Tetrahedron and Cube Patterns

For use with Investigation 12

Tetrahedron (four-faced polyhedron)

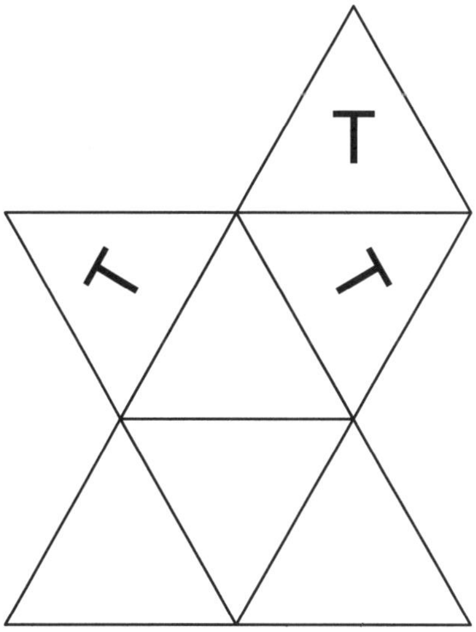

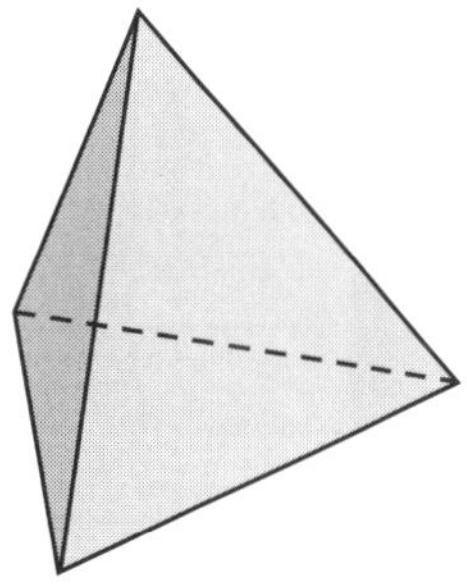

Cube

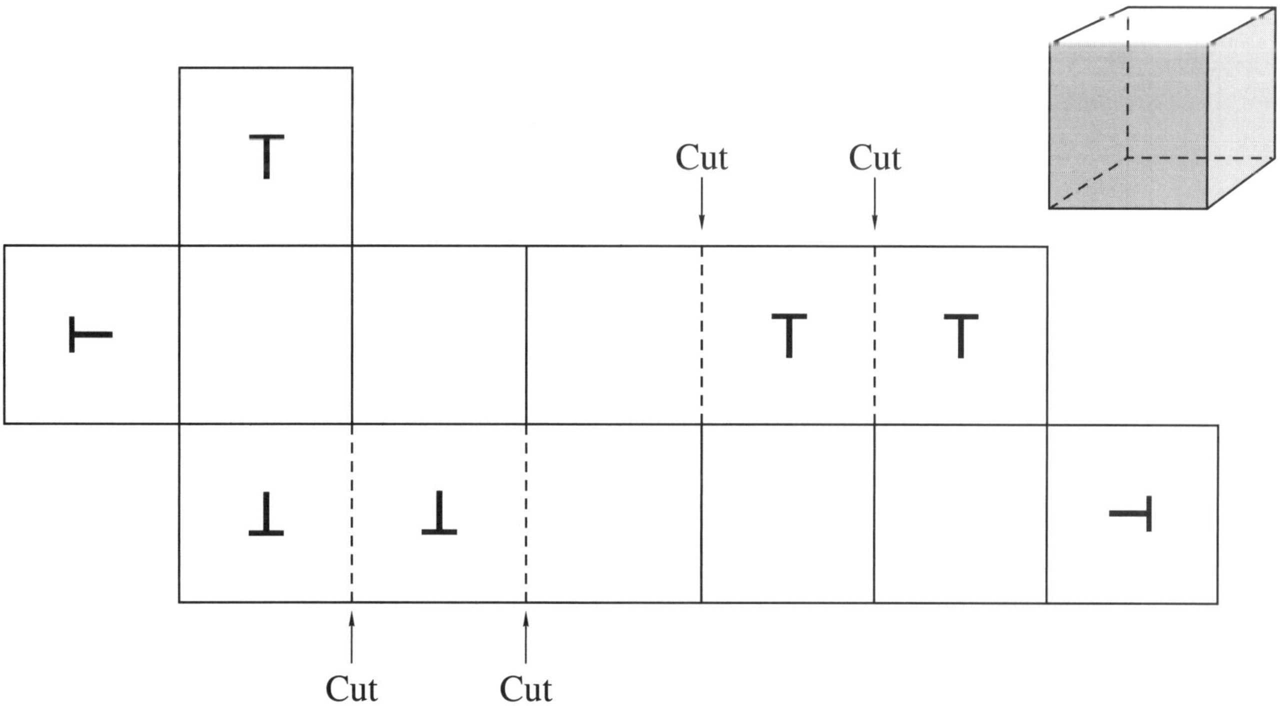

ACTIVITY MASTER

15 Octahedron and Dodecahedron Patterns

For use with Investigation 12

Octahedron

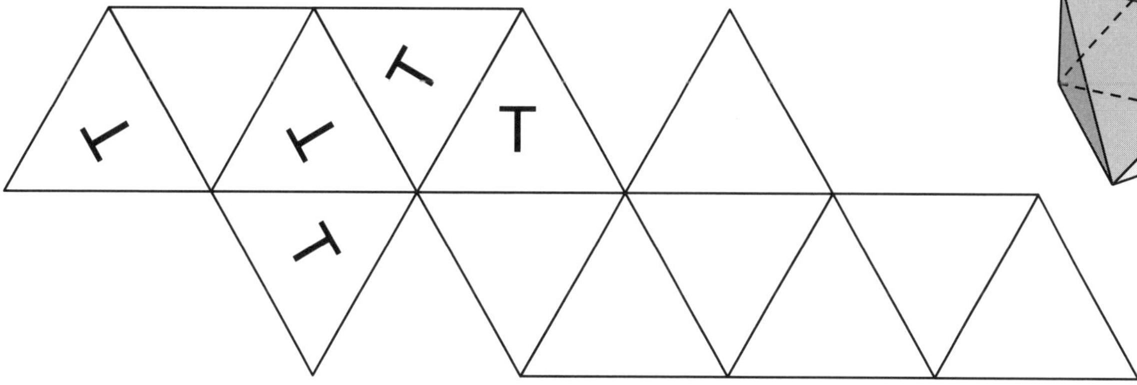

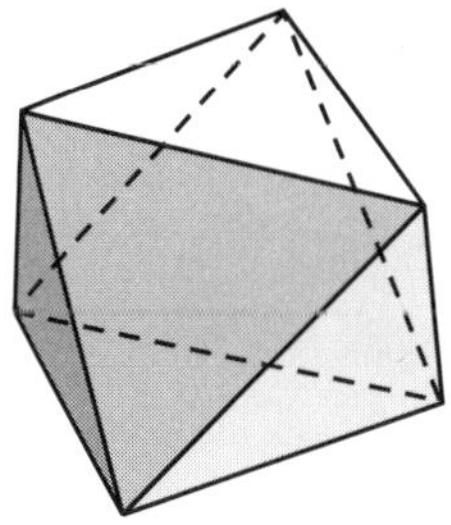

Dodecahedron

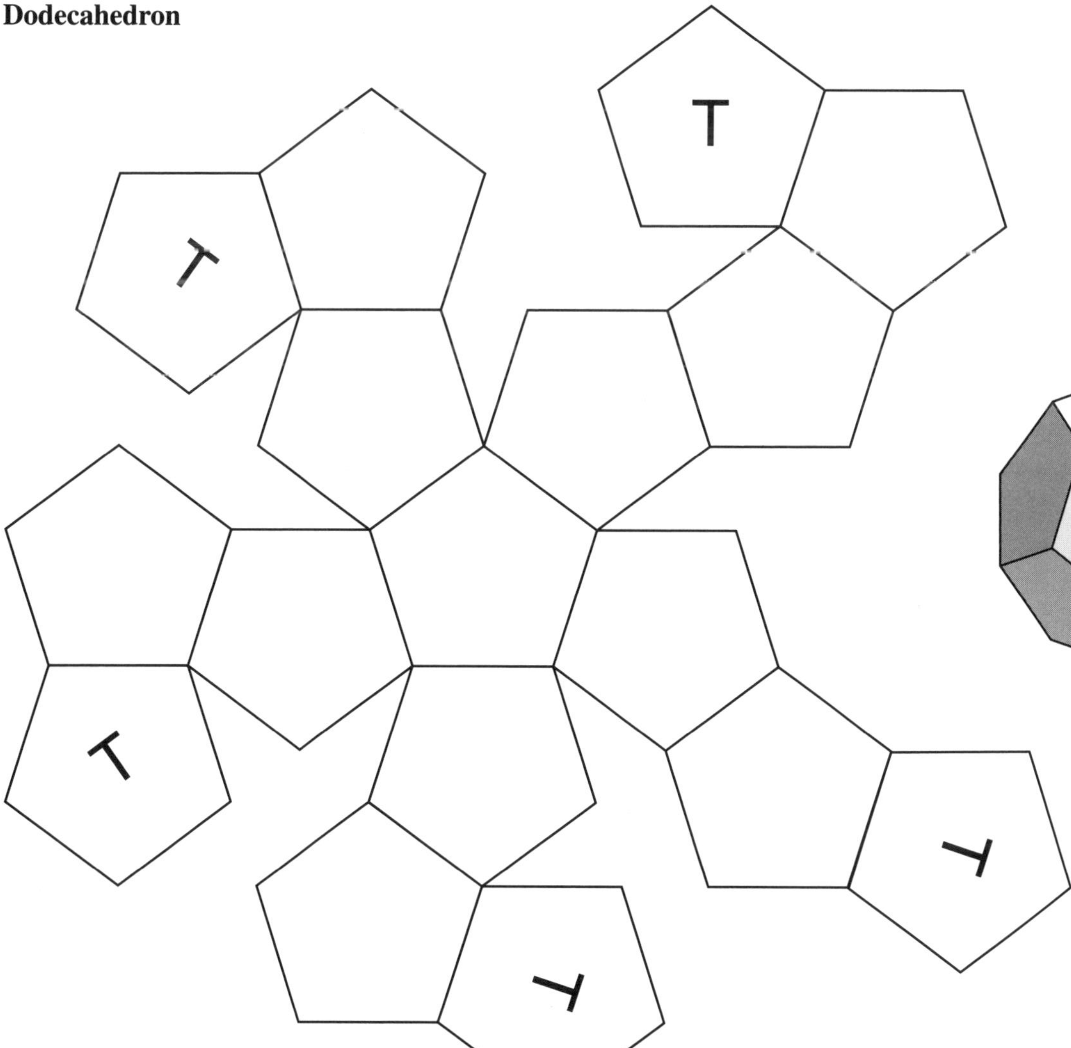

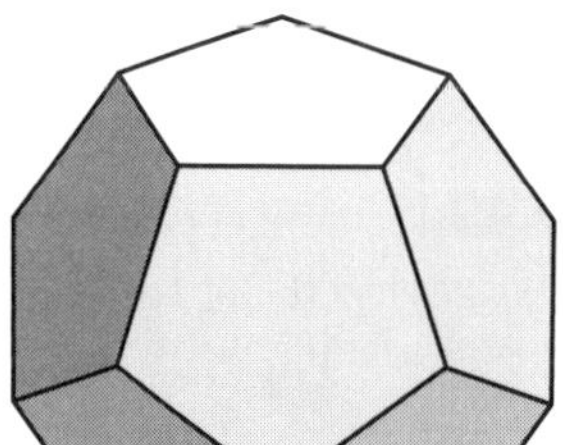

16

Icosahedron Pattern

For use with Investigation 12

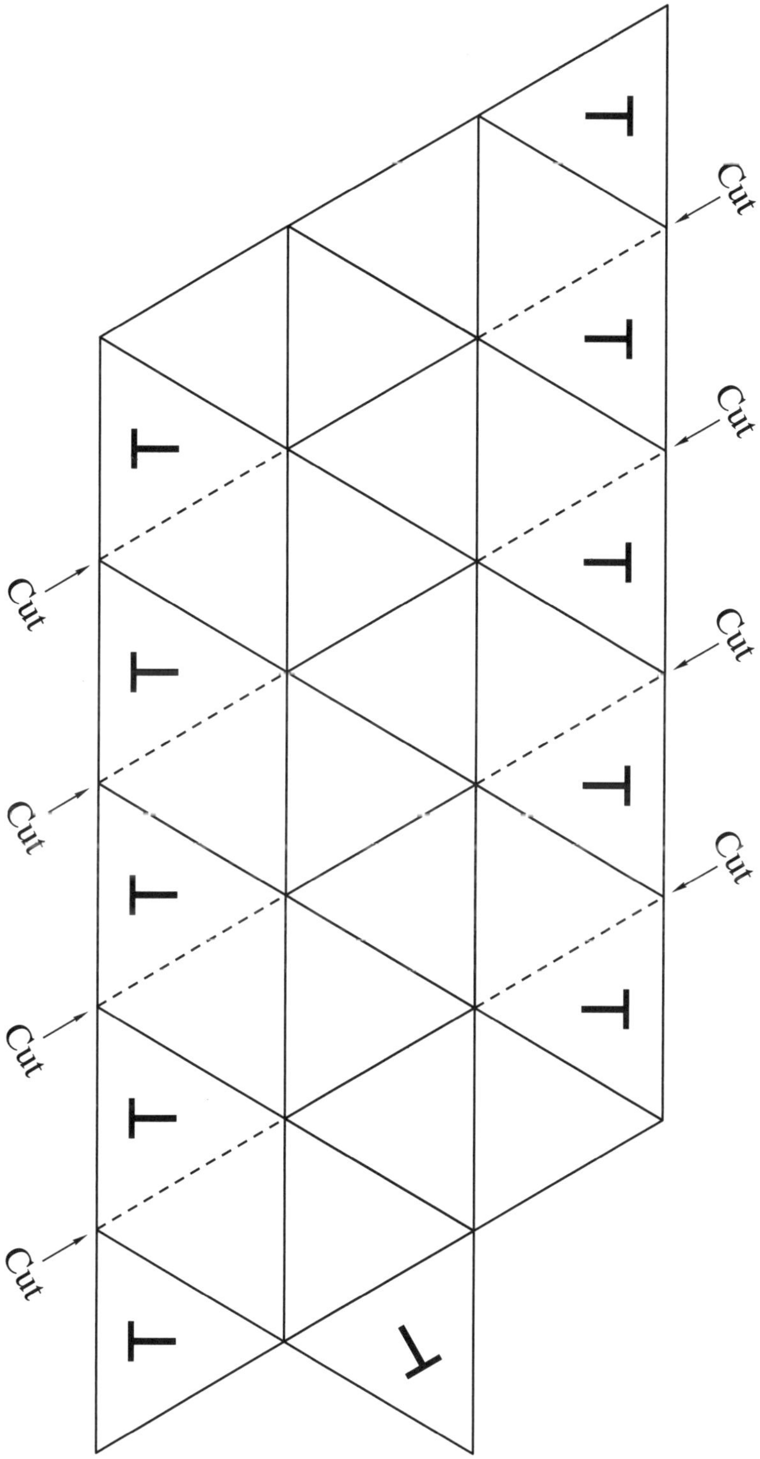

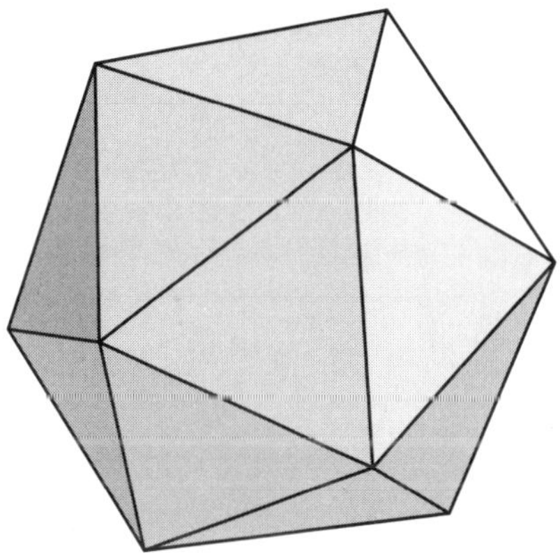

Answer Forms

Answer Form A: Lesson Worksheet
A single-sided master with check boxes for daily activities, answer blanks for warm-up activities, and partitions for recording the solutions to twelve practice problems

Answer Form B: Mixed Practice Solutions
A double-sided master with a grid background and partitions for recording the solutions to thirty problems

Answer Form C: Mixed Practice Solutions
A double-sided master with a plain, white background and partitions for recording the solutions to thirty problems

Answer Form D: Test Solutions
A double-sided master with a grid background and partitions for recording the solutions to twenty problems

Answer Form E: Test Solutions
A double-sided master with a plain, white background and partitions for recording the solutions to twenty problems

Answer Form F: Test Solutions
A single-sided master with partitions for recording the solutions to twenty problems and a separate answer column on the right-hand side

A

Lesson Worksheet

Show all necessary work. Please be neat.

Name ______________________

Date ______________________

Lesson ______________________

Warm-Up

☐ Facts Practice
☐ Mental Math
☐ Problem Solving

Review

☐ Homework Check
☐ Error Correction

Instruction

☐ Lesson
☐ Lesson Practice
☐ Mixed Practice

Facts Practice

Test:	Time:	Score:

Mental Math

a.	**b.**	**c.**	**d.**	**e.**	**f.**
g.	**h.**	**i.**	**j.**	**k.**	**l.**

Problem Solving

Strategies:
(Check any you use.)

☐ Make a chart, graph, or list.
☐ Guess and check (trial and error).
☐ Use logical reasoning.
☐ Act it out.
☐ Draw a diagram.
☐ Make it simpler.
☐ Draw a picture.
☐ Work backward.
☐ Find a pattern.

Lesson Practice

a.	**b.**	**c.**
d.	**e.**	**f.**
g.	**h.**	**i.**
j.	**k.**	**l.**

ANSWER FORM

E

Test Solutions
Show your work on this paper.
Do not write on the test.

Name ______________________

Date ______________________

Test __________ Score __________

	2.
	4.
	6.
	8.
	10.

11.

12.

13.

14.

15.

16.

17.

18.

19.

20.

ANSWER FORM

F

Test Solutions

Show your work on this paper.
Do not write on the test.

Name ______________________

Date ______________________ Test ________

2.	3.	4.	
			1.
			2.
			3.
			4.
6.	7.	8.	5.
			6.
			7.
			8.
10.	11.	12.	9.
			10.
			11.
			12.
14.	15.	16.	13.
			14.
			15.
			16.
18.	19.	20.	17.
			18.
			19.
			20.

Recording Forms

Recording Form A: Scorecard

Give one to each student so he or she [illegible] ments are due and record scores on assignments. You may also ask stude[illegible]nts to initial beside each score.

Recording Form B: Facts Practice

Give one to each student so he or she can monitor performances on Facts Practice Tests throughout the year.

Recording Form C: Test Analysis

Complete one form after every test to determine which concepts are causing students the most trouble. Remediate the problems with additional instruction or practice if necessary.

Recording Form D: Individual Test Analysis

Complete a separate form after every test for each student who scored below 80%. The form allows you to identify a student's misunderstandings and plan remediation activities.

A Scorecard

Name ______________________

Date Due	Lesson or Test	Score	Parent's Initials	Date Due	Lesson or Test	Score	Parent's Initials	Date Due	Lesson or Test	Score	Parent's Initials

RECORDING FORM

B Facts Practice

Name ______________________

TEST	# POSSIBLE	TIME AND SCORE (time / # correct)
A 64 Addition Facts B 100 Addition Facts	64 100	A A A A A A B B B B
C 100 Subtraction Facts	100	
D 64 Multiplication Facts	64	
E 100 Multiplication Facts F 90 Division Facts	100 90	E E E E F F F F F
G 30 Fractions to Reduce	30	
H 72 Multiplication and Division Facts	72	
I 28 Improper Fractions to Simplify	28	
J 24 Mixed Numbers to Write as Improper Fractions	24	
K Linear Measurement	41	
L Liquid Measurement	32	
M 24 Percent-Fraction-Decimal Equivalents	24	
N Measurement Facts	52	

RECORDING FORM

C Test Analysis

Section ______________________

Date ______________________

Number of students ________ Test ________

Problem	Type of problem	Number of students who missed the problem: Form A Tally	Form B Tally	TOTAL
1.				
2.				
3.				
4.				
5.				
6.				
7.				
8.				
9.				
10.				
11.				
12.				
13.				
14.				
15.				
16.				
17.				
18.				
19.				
20.				

RECORDING FORM

D Individual Test Analysis

Name ______________________

Date ______________________

Number of problems missed _______ Test _______

Write the number of each test problem missed.

Analyze the student's work and interview the student to determine why each problem was missed. Write each problem number under "yes" or "no" as appropriate for each question.

	YES					NO				
1. Was the problem copied correctly?										
2. Was a careless mistake made?										
3. Was a concept or skill misunderstood? (Specify below.)										

With the student, identify concepts and skills that need to be learned. Prescribe remediation exercises to enable the student to understand the concepts and skills.

Answers and Solutions

Facts Practice Answers

The pages in this section reproduce the Facts Practice Tests with the answers printed in boldface.

Test Answers

This section contains the answers to the Pretest and Posttest, weekly tests, and the Final Exam.

Test-Day Activity Masters Answers

This section contains the answers to all Test-Day Activity Master worksheets.

Activity Masters Answers

This section contains the answers to all Activity Master worksheets.

Test Solutions

This section contains the solutions to the Pretest and Posttest, the weekly tests, and the Final Exam. Though the solutions are representative of students' work, please keep in mind that many problems have more than one correct solution. The final answers are set in boldface for easy identification.

A

64 Addition Facts

Add.

7 + 2 **9**	9 + 4 **13**	2 + 8 **10**	6 + 5 **11**	4 + 4 **8**	3 + 9 **12**	8 + 4 **12**	5 + 7 **12**
9 + 7 **16**	4 + 7 **11**	7 + 5 **12**	5 + 4 **9**	3 + 4 **7**	6 + 8 **14**	2 + 5 **7**	8 + 8 **16**
6 + 3 **9**	2 + 9 **11**	7 + 8 **15**	8 + 3 **11**	5 + 9 **14**	3 + 6 **9**	9 + 9 **18**	4 + 9 **13**
5 + 8 **13**	9 + 5 **14**	4 + 5 **9**	8 + 6 **14**	2 + 3 **5**	6 + 6 **12**	5 + 2 **7**	7 + 3 **10**
3 + 8 **11**	8 + 9 **17**	2 + 2 **4**	7 + 6 **13**	5 + 5 **10**	6 + 9 **15**	3 + 7 **10**	9 + 8 **17**
4 + 2 **6**	3 + 3 **6**	6 + 4 **10**	4 + 8 **12**	9 + 3 **12**	2 + 4 **6**	8 + 5 **13**	7 + 9 **16**
7 + 4 **11**	2 + 6 **8**	5 + 3 **8**	9 + 6 **15**	4 + 3 **7**	6 + 7 **13**	3 + 2 **5**	8 + 7 **15**
5 + 6 **11**	8 + 2 **10**	3 + 5 **8**	6 + 2 **8**	7 + 7 **14**	4 + 6 **10**	9 + 2 **11**	2 + 7 **9**

B

100 Addition Facts

Add.

3 + 2 **5**	8 + 3 **11**	2 + 1 **3**	5 + 6 **11**	2 + 9 **11**	4 + 8 **12**	8 + 0 **8**	3 + 9 **12**	1 + 0 **1**	6 + 3 **9**
7 + 3 **10**	1 + 6 **7**	4 + 7 **11**	0 + 3 **3**	6 + 4 **10**	5 + 5 **10**	3 + 1 **4**	7 + 2 **9**	8 + 5 **13**	2 + 5 **7**
4 + 0 **4**	5 + 7 **12**	1 + 1 **2**	5 + 4 **9**	2 + 8 **10**	7 + 1 **8**	4 + 6 **10**	0 + 2 **2**	6 + 5 **11**	4 + 9 **13**
8 + 6 **14**	0 + 4 **4**	5 + 8 **13**	7 + 4 **11**	1 + 7 **8**	6 + 6 **12**	4 + 1 **5**	8 + 2 **10**	2 + 4 **6**	6 + 0 **6**
9 + 1 **10**	8 + 8 **16**	2 + 2 **4**	4 + 5 **9**	6 + 2 **8**	0 + 0 **0**	5 + 9 **14**	3 + 3 **6**	8 + 1 **9**	2 + 7 **9**
4 + 4 **8**	7 + 5 **12**	0 + 1 **1**	8 + 7 **15**	3 + 4 **7**	7 + 9 **16**	1 + 2 **3**	6 + 7 **13**	0 + 8 **8**	9 + 2 **11**
0 + 9 **9**	8 + 9 **17**	7 + 6 **13**	1 + 3 **4**	6 + 8 **14**	2 + 0 **2**	8 + 4 **12**	3 + 5 **8**	9 + 8 **17**	5 + 0 **5**
9 + 3 **12**	2 + 6 **8**	3 + 0 **3**	6 + 1 **7**	3 + 6 **9**	5 + 2 **7**	0 + 5 **5**	6 + 9 **15**	1 + 8 **9**	9 + 6 **15**
4 + 3 **7**	9 + 9 **18**	0 + 7 **7**	9 + 4 **13**	7 + 7 **14**	1 + 4 **5**	3 + 7 **10**	7 + 0 **7**	2 + 3 **5**	5 + 1 **6**
9 + 5 **14**	1 + 5 **6**	9 + 0 **9**	3 + 8 **11**	1 + 9 **10**	5 + 3 **8**	4 + 2 **6**	9 + 7 **16**	0 + 6 **6**	7 + 8 **15**

C

100 Subtraction Facts

Subtract.

16 − 9 **7**	7 − 1 **6**	18 − 9 **9**	11 − 3 **8**	13 − 7 **6**	8 − 2 **6**	11 − 5 **6**	5 − 0 **5**	17 − 9 **8**	6 − 1 **5**
10 − 9 **1**	6 − 2 **4**	13 − 4 **9**	4 − 0 **4**	10 − 5 **5**	5 − 1 **4**	10 − 3 **7**	12 − 6 **6**	10 − 1 **9**	6 − 4 **2**
7 − 2 **5**	14 − 7 **7**	8 − 1 **7**	11 − 6 **5**	3 − 3 **0**	16 − 7 **9**	5 − 2 **3**	12 − 4 **8**	3 − 0 **3**	11 − 7 **4**
17 − 8 **9**	6 − 0 **6**	10 − 6 **4**	4 − 1 **3**	9 − 5 **4**	9 − 0 **9**	5 − 4 **1**	12 − 5 **7**	4 − 2 **2**	9 − 3 **6**
12 − 3 **9**	16 − 8 **8**	9 − 1 **8**	15 − 6 **9**	11 − 4 **7**	13 − 5 **8**	1 − 0 **1**	8 − 5 **3**	9 − 6 **3**	11 − 2 **9**
7 − 0 **7**	10 − 8 **2**	6 − 3 **3**	14 − 5 **9**	3 − 1 **2**	8 − 6 **2**	4 − 4 **0**	11 − 8 **3**	3 − 2 **1**	15 − 9 **6**
13 − 8 **5**	7 − 4 **3**	10 − 7 **3**	0 − 0 **0**	12 − 8 **4**	5 − 5 **0**	4 − 3 **1**	8 − 7 **1**	7 − 3 **4**	7 − 6 **1**
5 − 3 **2**	7 − 5 **2**	2 − 1 **1**	6 − 6 **0**	8 − 4 **4**	2 − 2 **0**	13 − 6 **7**	15 − 8 **7**	2 − 0 **2**	13 − 9 **4**
1 − 1 **0**	11 − 9 **2**	10 − 4 **6**	9 − 2 **7**	14 − 6 **8**	8 − 0 **8**	9 − 4 **5**	10 − 2 **8**	6 − 5 **1**	8 − 3 **5**
7 − 7 **0**	14 − 8 **6**	12 − 9 **3**	9 − 8 **1**	12 − 7 **5**	9 − 9 **0**	15 − 7 **8**	8 − 8 **0**	14 − 9 **5**	9 − 7 **2**

D

64 Multiplication Facts

Multiply.

5 × 6 **30**	4 × 3 **12**	9 × 8 **72**	7 × 5 **35**	2 × 9 **18**	8 × 4 **32**	9 × 3 **27**	6 × 9 **54**
9 × 4 **36**	2 × 5 **10**	7 × 6 **42**	4 × 8 **32**	7 × 9 **63**	5 × 4 **20**	3 × 2 **6**	9 × 7 **63**
3 × 7 **21**	8 × 5 **40**	6 × 2 **12**	5 × 5 **25**	3 × 5 **15**	2 × 4 **8**	7 × 7 **49**	8 × 9 **72**
6 × 4 **24**	2 × 8 **16**	4 × 4 **16**	8 × 2 **16**	3 × 9 **27**	6 × 6 **36**	9 × 9 **81**	5 × 3 **15**
4 × 6 **24**	8 × 8 **64**	5 × 7 **35**	6 × 3 **18**	2 × 2 **4**	7 × 4 **28**	3 × 8 **24**	8 × 6 **48**
2 × 6 **12**	5 × 9 **45**	3 × 3 **9**	9 × 2 **18**	6 × 7 **42**	4 × 5 **20**	7 × 2 **14**	9 × 6 **54**
5 × 2 **10**	7 × 8 **56**	2 × 3 **6**	6 × 8 **48**	4 × 7 **28**	9 × 5 **45**	3 × 6 **18**	8 × 7 **56**
3 × 4 **12**	7 × 3 **21**	5 × 8 **40**	4 × 2 **8**	8 × 3 **24**	2 × 7 **14**	6 × 5 **30**	4 × 9 **36**

E

100 Multiplication Facts

Multiply.

9 × 9 **81**	3 × 5 **15**	8 × 5 **40**	2 × 6 **12**	4 × 7 **28**	0 × 3 **0**	7 × 2 **14**	1 × 5 **5**	7 × 8 **56**	4 × 0 **0**
3 × 4 **12**	5 × 9 **45**	0 × 2 **0**	7 × 3 **21**	4 × 1 **4**	2 × 7 **14**	6 × 3 **18**	5 × 4 **20**	1 × 0 **0**	9 × 2 **18**
1 × 1 **1**	9 × 0 **0**	2 × 8 **16**	6 × 4 **24**	0 × 7 **0**	8 × 1 **8**	3 × 3 **9**	4 × 8 **32**	9 × 3 **27**	2 × 0 **0**
4 × 9 **36**	7 × 0 **0**	1 × 2 **2**	8 × 4 **32**	6 × 5 **30**	2 × 9 **18**	9 × 4 **36**	0 × 1 **0**	7 × 4 **28**	5 × 8 **40**
0 × 8 **0**	4 × 2 **8**	9 × 8 **72**	3 × 6 **18**	5 × 5 **25**	1 × 6 **6**	5 × 0 **0**	6 × 6 **36**	2 × 1 **2**	7 × 9 **63**
9 × 1 **9**	2 × 2 **4**	5 × 1 **5**	4 × 3 **12**	0 × 0 **0**	8 × 9 **72**	3 × 7 **21**	9 × 7 **63**	1 × 7 **7**	6 × 0 **0**
5 × 6 **30**	7 × 5 **35**	3 × 0 **0**	8 × 8 **64**	1 × 3 **3**	8 × 3 **24**	5 × 2 **10**	0 × 4 **0**	9 × 5 **45**	6 × 7 **42**
2 × 3 **6**	8 × 6 **48**	0 × 5 **0**	6 × 1 **6**	3 × 8 **24**	7 × 6 **42**	1 × 8 **8**	9 × 6 **54**	4 × 4 **16**	5 × 3 **15**
7 × 7 **49**	1 × 4 **4**	6 × 2 **12**	4 × 5 **20**	2 × 4 **8**	8 × 0 **0**	3 × 1 **3**	6 × 8 **48**	0 × 9 **0**	8 × 7 **56**
3 × 2 **6**	4 × 6 **24**	1 × 9 **9**	5 × 7 **35**	8 × 2 **16**	0 × 6 **0**	7 × 1 **7**	2 × 5 **10**	6 × 9 **54**	3 × 9 **27**

F

90 Division Facts

Divide.

3 7) 21	**5** 2) 10	**7** 6) 42	**3** 1) 3	**6** 4) 24	**2** 3) 6	**6** 9) 54	**3** 6) 18	**0** 4) 0	**6** 5) 30
8 4) 32	**7** 8) 56	**0** 1) 0	**2** 6) 12	**6** 3) 18	**8** 9) 72	**3** 5) 15	**4** 2) 8	**6** 7) 42	**6** 6) 36
0 6) 0	**2** 5) 10	**1** 9) 9	**3** 2) 6	**9** 7) 63	**4** 4) 16	**6** 8) 48	**2** 1) 2	**7** 5) 35	**7** 3) 21
9 2) 18	**1** 6) 6	**5** 3) 15	**5** 8) 40	**0** 2) 0	**4** 5) 20	**3** 9) 27	**8** 1) 8	**1** 4) 4	**5** 7) 35
5 4) 20	**7** 9) 63	**4** 1) 4	**2** 7) 14	**1** 3) 3	**3** 8) 24	**0** 5) 0	**4** 6) 24	**1** 8) 8	**8** 2) 16
1 5) 5	**8** 8) 64	**0** 3) 0	**7** 4) 28	**7** 7) 49	**2** 2) 4	**9** 9) 81	**4** 3) 12	**5** 6) 30	**5** 1) 5
4 8) 32	**1** 1) 1	**4** 9) 36	**9** 3) 27	**7** 2) 14	**5** 5) 25	**8** 6) 48	**0** 8) 0	**4** 7) 28	**9** 4) 36
6 2) 12	**9** 5) 45	**7** 1) 7	**2** 4) 8	**0** 7) 0	**2** 8) 16	**8** 3) 24	**5** 9) 45	**9** 1) 9	**9** 6) 54
8 7) 56	**0** 9) 0	**9** 8) 72	**1** 2) 2	**8** 5) 40	**3** 3) 9	**2** 9) 18	**6** 1) 6	**3** 4) 12	**1** 7) 7

G

30 Fractions to Reduce

Cut copies into two tests.

Test G—30 Fractions to Reduce

Name ____________ Time ________

Reduce each fraction to lowest terms.

$\frac{2}{8} = \mathbf{\frac{1}{4}}$	$\frac{4}{6} = \mathbf{\frac{2}{3}}$	$\frac{6}{10} = \mathbf{\frac{3}{5}}$	$\frac{2}{4} = \mathbf{\frac{1}{2}}$	$\frac{6}{16} = \mathbf{\frac{3}{8}}$
$\frac{5}{100} = \mathbf{\frac{1}{20}}$	$\frac{9}{12} = \mathbf{\frac{3}{4}}$	$\frac{14}{16} = \mathbf{\frac{7}{8}}$	$\frac{4}{10} = \mathbf{\frac{2}{5}}$	$\frac{4}{12} = \mathbf{\frac{1}{3}}$
$\frac{2}{10} = \mathbf{\frac{1}{5}}$	$\frac{3}{6} = \mathbf{\frac{1}{2}}$	$\frac{25}{100} = \mathbf{\frac{1}{4}}$	$\frac{3}{12} = \mathbf{\frac{1}{4}}$	$\frac{4}{16} = \mathbf{\frac{1}{4}}$
$\frac{3}{9} = \mathbf{\frac{1}{3}}$	$\frac{10}{16} = \mathbf{\frac{5}{8}}$	$\frac{6}{9} = \mathbf{\frac{2}{3}}$	$\frac{4}{8} = \mathbf{\frac{1}{2}}$	$\frac{2}{12} = \mathbf{\frac{1}{6}}$
$\frac{6}{12} = \mathbf{\frac{1}{2}}$	$\frac{2}{16} = \mathbf{\frac{1}{8}}$	$\frac{8}{10} = \mathbf{\frac{4}{5}}$	$\frac{2}{6} = \mathbf{\frac{1}{3}}$	$\frac{75}{100} = \mathbf{\frac{3}{4}}$
$\frac{12}{16} = \mathbf{\frac{3}{4}}$	$\frac{8}{12} = \mathbf{\frac{2}{3}}$	$\frac{6}{8} = \mathbf{\frac{3}{4}}$	$\frac{10}{12} = \mathbf{\frac{5}{6}}$	$\frac{5}{10} = \mathbf{\frac{1}{2}}$

Name ____________ Time ________

Reduce each fraction to lowest terms.

$\frac{2}{8} = \mathbf{\frac{1}{4}}$	$\frac{4}{6} = \mathbf{\frac{2}{3}}$	$\frac{6}{10} = \mathbf{\frac{3}{5}}$	$\frac{2}{4} = \mathbf{\frac{1}{2}}$	$\frac{6}{16} = \mathbf{\frac{3}{8}}$
$\frac{5}{100} = \mathbf{\frac{1}{20}}$	$\frac{9}{12} = \mathbf{\frac{3}{4}}$	$\frac{14}{16} = \mathbf{\frac{7}{8}}$	$\frac{4}{10} = \mathbf{\frac{2}{5}}$	$\frac{4}{12} = \mathbf{\frac{1}{3}}$
$\frac{2}{10} = \mathbf{\frac{1}{5}}$	$\frac{3}{6} = \mathbf{\frac{1}{2}}$	$\frac{25}{100} = \mathbf{\frac{1}{4}}$	$\frac{3}{12} = \mathbf{\frac{1}{4}}$	$\frac{4}{16} = \mathbf{\frac{1}{4}}$
$\frac{3}{9} = \mathbf{\frac{1}{3}}$	$\frac{10}{16} = \mathbf{\frac{5}{8}}$	$\frac{6}{9} = \mathbf{\frac{2}{3}}$	$\frac{4}{8} = \mathbf{\frac{1}{2}}$	$\frac{2}{12} = \mathbf{\frac{1}{6}}$
$\frac{6}{12} = \mathbf{\frac{1}{2}}$	$\frac{2}{16} = \mathbf{\frac{1}{8}}$	$\frac{8}{10} = \mathbf{\frac{4}{5}}$	$\frac{2}{6} = \mathbf{\frac{1}{3}}$	$\frac{75}{100} = \mathbf{\frac{3}{4}}$
$\frac{12}{16} = \mathbf{\frac{3}{4}}$	$\frac{8}{12} = \mathbf{\frac{2}{3}}$	$\frac{6}{8} = \mathbf{\frac{3}{4}}$	$\frac{10}{12} = \mathbf{\frac{5}{6}}$	$\frac{5}{10} = \mathbf{\frac{1}{2}}$

H

72 Multiplication and Division Facts

Multiply or divide as indicated.

5 × 9 = **45**	6) 36 = **6**	4 × 7 = **28**	8) 40 = **5**	10 × 3 = **30**	5) 20 = **4**	2 × 7 = **14**	9) 27 = **3**
9) 81 = **9**	5 × 7 = **35**	6) 24 = **4**	2 × 3 = **6**	7) 42 = **6**	7 × 9 = **63**	5) 10 = **2**	4 × 5 = **20**
2 × 9 = **18**	7) 28 = **4**	6 × 9 = **54**	4) 8 = **2**	3 × 3 = **9**	8) 24 = **3**	10 × 4 = **40**	6) 12 = **2**
9) 63 = **7**	2 × 5 = **10**	8) 64 = **8**	7 × 6 = **42**	10) 100 = **10**	3 × 5 = **15**	9) 36 = **4**	10 × 8 = **80**
4 × 8 = **32**	5) 25 = **5**	3 × 8 = **24**	4) 16 = **4**	10 × 5 = **50**	7) 14 = **2**	2 × 2 = **4**	8) 48 = **6**
9) 54 = **6**	3 × 6 = **18**	5) 15 = **3**	2 × 8 = **16**	7) 49 = **7**	4 × 6 = **24**	6) 30 = **5**	5 × 8 = **40**
10 × 6 = **60**	4) 12 = **3**	7 × 8 = **56**	6) 18 = **3**	3 × 4 = **12**	7) 35 = **5**	5 × 6 = **30**	8) 32 = **4**
8) 56 = **7**	2 × 4 = **8**	8) 16 = **2**	6 × 8 = **48**	7) 21 = **3**	8 × 9 = **72**	3) 6 = **2**	3 × 9 = **27**
10 × 7 = **70**	9) 18 = **2**	3 × 7 = **21**	9) 45 = **5**	2 × 6 = **12**	9) 72 = **8**	4 × 9 = **36**	9) 90 = **10**

I

28 Improper Fractions to Simplify

Cut copies into two tests.

Test I—28 Improper Fractions to Simplify

Name ________ Time ________

Write each improper fraction as a mixed number or a whole number.

$\frac{5}{4} = \mathbf{1\frac{1}{4}}$	$\frac{16}{12} = \mathbf{1\frac{1}{3}}$	$\frac{12}{8} = \mathbf{1\frac{1}{2}}$	$\frac{8}{6} = \mathbf{1\frac{1}{3}}$
$\frac{12}{6} = \mathbf{2}$	$\frac{12}{10} = \mathbf{1\frac{1}{5}}$	$\frac{6}{4} = \mathbf{1\frac{1}{2}}$	$\frac{20}{12} = \mathbf{1\frac{2}{3}}$
$\frac{5}{3} = \mathbf{1\frac{2}{3}}$	$\frac{10}{8} = \mathbf{1\frac{1}{4}}$	$\frac{25}{10} = \mathbf{2\frac{1}{2}}$	$\frac{10}{3} = \mathbf{3\frac{1}{3}}$
$\frac{15}{10} = \mathbf{1\frac{1}{2}}$	$\frac{3}{2} = \mathbf{1\frac{1}{2}}$	$\frac{9}{6} = \mathbf{1\frac{1}{2}}$	$\frac{7}{4} = \mathbf{1\frac{3}{4}}$
$\frac{18}{12} = \mathbf{1\frac{1}{2}}$	$\frac{8}{3} = \mathbf{2\frac{2}{3}}$	$\frac{15}{6} = \mathbf{2\frac{1}{2}}$	$\frac{14}{4} = \mathbf{3\frac{1}{2}}$
$\frac{8}{4} = \mathbf{2}$	$\frac{10}{6} = \mathbf{1\frac{2}{3}}$	$\frac{5}{2} = \mathbf{2\frac{1}{2}}$	$\frac{21}{12} = \mathbf{1\frac{3}{4}}$
$\frac{15}{12} = \mathbf{1\frac{1}{4}}$	$\frac{10}{4} = \mathbf{2\frac{1}{2}}$	$\frac{15}{8} = \mathbf{1\frac{7}{8}}$	$\frac{4}{3} = \mathbf{1\frac{1}{3}}$

Test I—28 Improper Fractions to Simplify

Name ________ Time ________

Write each improper fraction as a mixed number or a whole number.

$\frac{5}{4} = \mathbf{1\frac{1}{4}}$	$\frac{16}{12} = \mathbf{1\frac{1}{3}}$	$\frac{12}{8} = \mathbf{1\frac{1}{2}}$	$\frac{8}{6} = \mathbf{1\frac{1}{3}}$
$\frac{12}{6} = \mathbf{2}$	$\frac{12}{10} = \mathbf{1\frac{1}{5}}$	$\frac{6}{4} = \mathbf{1\frac{1}{2}}$	$\frac{20}{12} = \mathbf{1\frac{2}{3}}$
$\frac{5}{3} = \mathbf{1\frac{2}{3}}$	$\frac{10}{8} = \mathbf{1\frac{1}{4}}$	$\frac{25}{10} = \mathbf{2\frac{1}{2}}$	$\frac{10}{3} = \mathbf{3\frac{1}{3}}$
$\frac{15}{10} = \mathbf{1\frac{1}{2}}$	$\frac{3}{2} = \mathbf{1\frac{1}{2}}$	$\frac{9}{6} = \mathbf{1\frac{1}{2}}$	$\frac{7}{4} = \mathbf{1\frac{3}{4}}$
$\frac{18}{12} = \mathbf{1\frac{1}{2}}$	$\frac{8}{3} = \mathbf{2\frac{2}{3}}$	$\frac{15}{6} = \mathbf{2\frac{1}{2}}$	$\frac{14}{4} = \mathbf{3\frac{1}{2}}$
$\frac{8}{4} = \mathbf{2}$	$\frac{10}{6} = \mathbf{1\frac{2}{3}}$	$\frac{5}{2} = \mathbf{2\frac{1}{2}}$	$\frac{21}{12} = \mathbf{1\frac{3}{4}}$
$\frac{15}{12} = \mathbf{1\frac{1}{4}}$	$\frac{10}{4} = \mathbf{2\frac{1}{2}}$	$\frac{15}{8} = \mathbf{1\frac{7}{8}}$	$\frac{4}{3} = \mathbf{1\frac{1}{3}}$

J

24 Mixed Numbers to Write as Improper Fractions

Cut copies into two tests.

Test J—24 Mixed Numbers to Write as Improper Fractions

Name ________________ Time ________

Write each mixed number as an improper fraction.

$2\frac{1}{2} = \mathbf{\frac{5}{2}}$	$2\frac{2}{5} = \mathbf{\frac{12}{5}}$	$1\frac{3}{4} = \mathbf{\frac{7}{4}}$	$2\frac{3}{4} = \mathbf{\frac{11}{4}}$
$2\frac{1}{8} = \mathbf{\frac{17}{8}}$	$1\frac{2}{3} = \mathbf{\frac{5}{3}}$	$10\frac{1}{2} = \mathbf{\frac{21}{2}}$	$2\frac{1}{3} = \mathbf{\frac{7}{3}}$
$3\frac{1}{2} = \mathbf{\frac{7}{2}}$	$1\frac{5}{6} = \mathbf{\frac{11}{6}}$	$2\frac{1}{4} = \mathbf{\frac{9}{4}}$	$1\frac{1}{8} = \mathbf{\frac{9}{8}}$
$5\frac{1}{2} = \mathbf{\frac{11}{2}}$	$1\frac{3}{8} = \mathbf{\frac{11}{8}}$	$5\frac{1}{3} = \mathbf{\frac{16}{3}}$	$3\frac{1}{4} = \mathbf{\frac{13}{4}}$
$4\frac{1}{2} = \mathbf{\frac{9}{2}}$	$1\frac{7}{8} = \mathbf{\frac{15}{8}}$	$2\frac{2}{3} = \mathbf{\frac{8}{3}}$	$3\frac{3}{10} = \mathbf{\frac{33}{10}}$
$1\frac{5}{8} = \mathbf{\frac{13}{8}}$	$3\frac{3}{4} = \mathbf{\frac{15}{4}}$	$2\frac{3}{8} = \mathbf{\frac{19}{8}}$	$7\frac{1}{2} = \mathbf{\frac{15}{2}}$

Test J—24 Mixed Numbers to Write as Improper Fractions

Name ________________ Time ________

Write each mixed number as an improper fraction.

$2\frac{1}{2} = \mathbf{\frac{5}{2}}$	$2\frac{2}{5} = \mathbf{\frac{12}{5}}$	$1\frac{3}{4} = \mathbf{\frac{7}{4}}$	$2\frac{3}{4} = \mathbf{\frac{11}{4}}$
$2\frac{1}{8} = \mathbf{\frac{17}{8}}$	$1\frac{2}{3} = \mathbf{\frac{5}{3}}$	$10\frac{1}{2} = \mathbf{\frac{21}{2}}$	$2\frac{1}{3} = \mathbf{\frac{7}{3}}$
$3\frac{1}{2} = \mathbf{\frac{7}{2}}$	$1\frac{5}{6} = \mathbf{\frac{11}{6}}$	$2\frac{1}{4} = \mathbf{\frac{9}{4}}$	$1\frac{1}{8} = \mathbf{\frac{9}{8}}$
$5\frac{1}{2} = \mathbf{\frac{11}{2}}$	$1\frac{3}{8} = \mathbf{\frac{11}{8}}$	$5\frac{1}{3} = \mathbf{\frac{16}{3}}$	$3\frac{1}{4} = \mathbf{\frac{13}{4}}$
$4\frac{1}{2} = \mathbf{\frac{9}{2}}$	$1\frac{7}{8} = \mathbf{\frac{15}{8}}$	$2\frac{2}{3} = \mathbf{\frac{8}{3}}$	$3\frac{3}{10} = \mathbf{\frac{33}{10}}$
$1\frac{5}{8} = \mathbf{\frac{13}{8}}$	$3\frac{3}{4} = \mathbf{\frac{15}{4}}$	$2\frac{3}{8} = \mathbf{\frac{19}{8}}$	$7\frac{1}{2} = \mathbf{\frac{15}{2}}$

K Linear Measurement

Write the abbreviation for each of the following units.

Metric Units

1. millimeter **mm**
2. centimeter **cm**
3. meter **m**
4. kilometer **km**

U.S. Customary Units

5. inch **in.**
6. foot **ft**
7. yard **yd**
8. mile **mi**

Complete each unit conversion.

Metric Conversions

9. 1 centimeter = **10** millimeters
10. 1 meter = **1000** millimeters
11. 1 meter = **100** centimeters
12. 1 kilometer = **1000** meters

U.S. Customary Conversions

13. 1 foot = **12** inches
14. 1 yard = **36** inches
15. 1 yard = **3** feet
16. 1 mile = **5280** feet
17. 1 mile = **1760** yards

Conversions between systems

18. 1 inch = **2.54** centimeters
19. 1 mile ≈ **1609** meters
20. 1 meter ≈ **39** inches
21. 1 kilometer ≈ $\mathbf{\frac{5}{8}}$ mile

Write an appropriate unit for each physical reference.

Metric Units

22. The thickness of a dime: **1 mm**
23. The width of a little finger: **1 cm**
24. The length of one BIG step: **1 m**

U.S. Customary Units

25. The width of two fingers: **1 in.**
26. The length of a man's shoe: **1 ft**
27. The length of one big step: **1 yd**

Arrange each set of units in order from shortest to longest.

28. m, cm, mm, km **mm**, **cm**, **m**, **km**
29. ft, mi, in., yd **in.**, **ft**, **yd**, **mi**

Find each equivalent measure.

30. 10 cm = **100** mm
31. 2 m = **200** cm or **2000** mm
32. 5 km = **5000** m
33. 2.5 cm = **25** mm
34. 1.5 m = **150** cm or **1500** mm
35. 7.5 km = **7500** m
36. $\frac{1}{2}$ ft = **6** in.
37. 2 ft = **24** in.
38. 3 ft = **36** in.
39. 2 yd = **6** ft
40. 10 yd = **30** ft
41. 100 yd = **300** ft

L Liquid Measurement

Write the abbreviation for each of the following units.

Metric Units

1. liter **L**
2. milliliter **mL**

U.S. Customary Units

3. ounce **oz**
4. pint **pt**
5. quart **qt**
6. gallon **gal**

Complete each unit conversion.

Metric Conversions

7. 1 liter = **1000** milliliters

U.S. Customary Conversions

8. 1 cup = **8** ounces
9. 1 pint = **16** ounces
10. 1 quart = **2** pints
11. 1 gallon = **4** quarts

Conversions between systems

12. 1 liter ≈ **1.06** quarts
13. 1 gallon ≈ **3.78** liters

Complete each statement.

14. One fourth of a dollar is a quarter, and one fourth of a gallon is a **quart**.
15. A two-liter bottle of soda is a little more than 2 **quarts** or $\frac{1}{2}$ of a **gallon**.
16. "A pint's a pound the world around" means a **pint** of water weighs about 1 pound.
17. In the United States, gasoline is sold by the **gallon**. In other countries, gasoline is sold by the **liter**. A gallon of gasoline is more than **3** liters but a little less than **4** liters.
18. 2 cups make a **pint**.
19. 2 pints make a **quart**.
20. 2 quarts make a **half gallon**.
21. 2 half gallons make a **gallon**.
22. A gallon of milk will fill **16** cups.
23. If you drink 8 cups of water each day, you drink a **half gallon** of water.

Find each equivalent measure.

24. 2 liters = **2000** milliliters
25. 2 liters ≈ **2 (or 2.12)** quarts
26. 3.78 liters = **3780** milliliters
27. 0.5 liters = **500** milliliters
28. $\frac{1}{2}$ pint (1 cup) = **8** ounces
29. 1 quart (2 pints) = **32** ounces
30. $\frac{1}{2}$ gallon = **2** quarts
31. 1 gallon = **8** pints
32. 2 gallons = **8** quarts

M

24 Percent-Fraction-Decimal Equivalents

Write each percent as a reduced fraction and a decimal.

Percent	Fraction	Decimal	Percent	Fraction	Decimal
40%	$\frac{2}{5}$	**0.4**	70%	$\frac{7}{10}$	**0.7**
5%	$\frac{1}{20}$	**0.05**	4%	$\frac{1}{25}$	**0.04**
80%	$\frac{4}{5}$	**0.8**	100%	**1**	**1**
2%	$\frac{1}{50}$	**0.02**	30%	$\frac{3}{10}$	**0.3**
3%	$\frac{3}{100}$	**0.03**	50%	$\frac{1}{2}$	**0.5**
20%	$\frac{1}{5}$	**0.2**	$12\frac{1}{2}\%$	$\frac{1}{8}$	**0.125**
25%	$\frac{1}{4}$	**0.25**	$37\frac{1}{2}\%$	$\frac{3}{8}$	**0.375**
60%	$\frac{3}{5}$	**0.6**	$62\frac{1}{2}\%$	$\frac{5}{8}$	**0.625**
1%	$\frac{1}{100}$	**0.01**	$87\frac{1}{2}\%$	$\frac{7}{8}$	**0.875**
90%	$\frac{9}{10}$	**0.9**	$33\frac{1}{3}\%$	$\frac{1}{3}$	Rounds to 0.333
75%	$\frac{3}{4}$	**0.75**	$66\frac{2}{3}\%$	$\frac{2}{3}$	Rounds to 0.667
10%	$\frac{1}{10}$	**0.1**	$16\frac{2}{3}\%$	$\frac{1}{6}$	Rounds to 0.167

N Measurement Facts

1. Draw a segment 1 cm long. **See student work.**
2. Draw a segment 1 in. long. **See student work.**
3. One inch is how many centimeters? **2.54 cm**
4. Which is longer, 1 km or 1 mi? **1 mi**
5. Which is longer, 1 km or $\frac{1}{2}$ mi? **1 km**
6. How many ounces are in a pound? **16 oz**
7. How many pounds are in a ton? **2000 lb**
8. A dollar bill has a mass of about one **gram**.
9. Your math book has a mass of about one **kilogram**.
10. On earth a kilogram mass weighs about **2.2** pounds.
11. A metric ton is **1000** kilograms.
12. On earth a metric ton weighs about **2200** pounds.
13. The earth rotates on its axis once in a **day**.
14. The earth revolves around the sun once in a **year**.

Abbreviations:

15. milligram **mg**
16. gram **g**
17. kilogram **kg**
18. **Celsius** C
19. ounce **oz**
20. pound **lb**
21. ton **tn**
22. **Fahrenheit** F

Equivalents:

23. 1 gram = **1000** milligrams
24. 1 kilogram = **1000** grams
25. $\frac{1}{2}$ ton = **1000** pounds
26. **365** days = a common year
27. **366** days = a leap year
28. **52** weeks ≈ a year
29. **10** years = a decade
30. **100** years = a century
31. **1000** years = a millennium

How many days are in

32. Jan. **31**
33. Feb. **28** or **29**
34. Mar. **31**
35. Apr. **30**
36. May **31**
37. June **30**
38. July **31**
39. Aug. **31**
40. Sept. **30**
41. Oct. **31**
42. Nov. **30**
43. Dec. **31**

Write the indicated temperatures.

44. Water boils **212** °F
45. **100** °C
46. Normal body temperature **98.6** °F
47. **37** °C

Cool room temperature **68** °F
48. **20** °C
49. Water freezes **32** °F
50. **0** °C

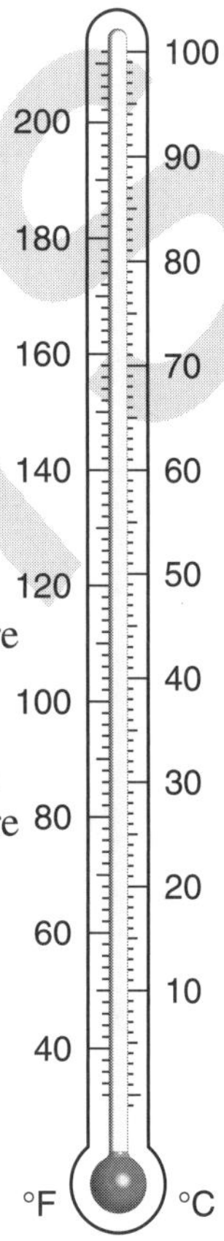

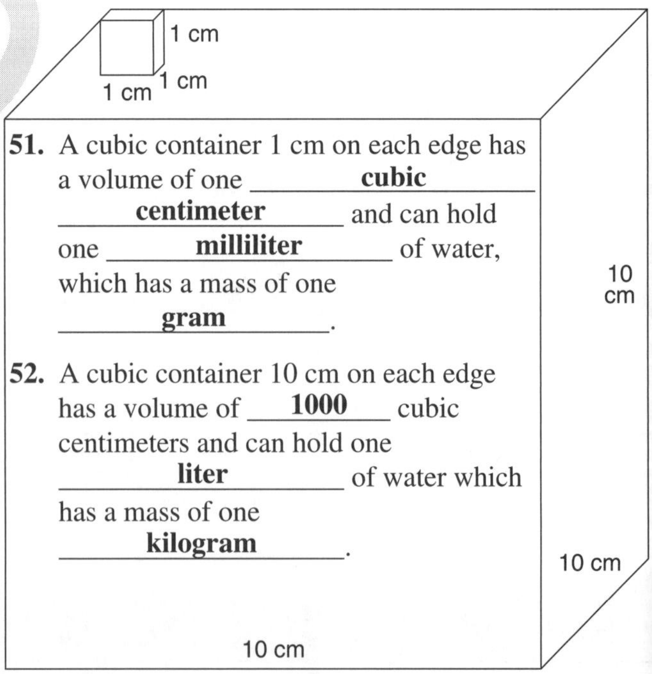

51. A cubic container 1 cm on each edge has a volume of one **cubic centimeter** and can hold one **milliliter** of water, which has a mass of one **gram**.

52. A cubic container 10 cm on each edge has a volume of **1000** cubic centimeters and can hold one **liter** of water which has a mass of one **kilogram**.

O Adding Integers

Simplify.

$(+3) + (-4) = \mathbf{-1}$	$(-2) + (+2) = \mathbf{0}$	$(-3) + (+4) = \mathbf{+1}$
$(+2) + (+2) = \mathbf{+4}$	$(-3) + (-4) = \mathbf{-7}$	$(-2) + (-2) = \mathbf{-4}$
$(+3) + (+4) = \mathbf{+7}$	$(+2) + (-2) = \mathbf{0}$	$(-5) + (+2) = \mathbf{-3}$
$(+5) + (-5) = \mathbf{0}$	$(-5) + (-2) = \mathbf{-7}$	$(+5) + (-2) = \mathbf{+3}$
$-6 + +4 = \mathbf{-2}$	$-5 + -5 = \mathbf{-10}$	$-6 + -4 = \mathbf{-10}$
$+5 + -5 = \mathbf{0}$	$+6 + -4 = \mathbf{+2}$	$+5 + +5 = \mathbf{+10}$
$+10 + -5 = \mathbf{+5}$	$-10 + +10 = \mathbf{0}$	$-10 + -5 = \mathbf{-15}$
$+10 + -10 = \mathbf{0}$	$-10 + +5 = \mathbf{-5}$	$-10 + -10 = \mathbf{-20}$
$6 + (-3) = \mathbf{+3}$	$5 + -8 = \mathbf{-3}$	$-6 + 4 = \mathbf{-2}$
$(-3) + -5 = \mathbf{-8}$	$6 + (+3) = \mathbf{+9}$	$-3 + 4 = \mathbf{+1}$
$-6 + (-5) = \mathbf{-11}$	$8 + (-2) = \mathbf{+6}$	$-3 + (-4) = \mathbf{-7}$
$-8 + (-8) = \mathbf{-16}$	$8 + (-8) = \mathbf{0}$	$-8 + 8 = \mathbf{0}$

P Adding and Subtracting Integers

Simplify.

(–3) – (+3) = **–6**	(–4) + (+6) = **+2**	5 + –2 = **+3**
(–7) + (–1) = **–8**	10 – (–10) = **+20**	(–3) + (+3) = **0**
10 – (–10) = **+20**	(–3) – (–3) = **0**	–5 + 2 = **–3**
(–4) – (+6) = **–10**	(–7) – (–1) = **–6**	4 + (–6) = **–2**
10 + (–8) = **+2**	(+7) – (–1) = **+8**	(+3) – (–3) = **+6**
(+7) – (+1) = **+6**	5 – –2 = **+7**	(–4) + (–6) = **–10**
(+3) – (+3) = **0**	–10 – 8 = **–18**	(–7) – (+1) = **–8**
10 + (–10) = **0**	(–7) + (+1) = **–6**	–5 – 2 = **–7**
–5 + –2 = **–7**	(+3) + (+3) = **+6**	–10 + (–8) = **–18**
–10 – (–8) = **–2**	(–4) – (–6) = **+2**	(+7) + (+1) = **+8**
(–3) + (–3) = **–6**	10 – (–8) = **+18**	(+3) + (–3) = **0**
4 – (–6) = **+10**	–5 – –2 = **–3**	(+7) + (–1) = **+6**

Q Multiplying and Dividing Integers

Simplify.

(–3)(–6) = **+18**	$\frac{-8}{-4}$ = **+2**	(–2)(+5) = **–10**	$\frac{+5}{-5}$ = **–1**
(+12) ÷ (+4) = **+3**	(–6)(+4) = **–24**	$\frac{-15}{5}$ = **–3**	(+5)(+2) = **+10**
(–2)(+8) = **–16**	$\frac{-5}{+5}$ = **–1**	(–6)(–4) = **+24**	$\frac{-8}{+4}$ = **–2**
$\frac{+5}{+5}$ = **+1**	(2)(+8) = **+16**	(+12) ÷ (–4) = **–3**	(3)(–6) = **–18**
(+3)(+6) = **+18**	$\frac{-15}{5}$ = **+3**	(–2)(–8) = **+16**	$\frac{+8}{4}$ = **–2**
(–12) ÷ (+4) = **–3**	(–5)(–2) = **+10**	$\frac{+15}{+5}$ = **+3**	(+6)(–4) = **–24**
(–3)(6) = **–18**	$\frac{+15}{-5}$ = **–3**	(+2)(–8) = **–16**	$\frac{-5}{-5}$ = **+1**
$\frac{+8}{+4}$ = **+2**	(+6)(+4) = **+24**	(–12) ÷ (–4) = **+3**	(–5)(+2) = **–10**

R

90 Division Facts

Divide.

20 ÷ 4 = **5**	21 ÷ 7 = **3**	0 ÷ 2 = **0**	27 ÷ 3 = **9**	8 ÷ 1 = **8**	54 ÷ 6 = **9**
15 ÷ 5 = **3**	6 ÷ 3 = **2**	28 ÷ 4 = **7**	18 ÷ 2 = **9**	24 ÷ 6 = **4**	9 ÷ 9 = **1**
56 ÷ 8 = **7**	0 ÷ 6 = **0**	21 ÷ 3 = **7**	1 ÷ 1 = **1**	25 ÷ 5 = **5**	12 ÷ 2 = **6**
5 ÷ 1 = **5**	45 ÷ 9 = **5**	16 ÷ 4 = **4**	30 ÷ 6 = **5**	9 ÷ 3 = **3**	14 ÷ 7 = **2**
0 ÷ 8 = **0**	6 ÷ 2 = **3**	24 ÷ 8 = **3**	10 ÷ 5 = **2**	81 ÷ 9 = **9**	24 ÷ 4 = **6**
16 ÷ 2 = **8**	30 ÷ 5 = **6**	0 ÷ 1 = **0**	28 ÷ 7 = **4**	4 ÷ 4 = **1**	40 ÷ 8 = **5**
3 ÷ 3 = **1**	18 ÷ 6 = **3**	63 ÷ 9 = **7**	40 ÷ 5 = **8**	10 ÷ 2 = **5**	36 ÷ 6 = **6**
32 ÷ 8 = **4**	12 ÷ 4 = **3**	18 ÷ 3 = **6**	35 ÷ 7 = **5**	8 ÷ 8 = **1**	2 ÷ 1 = **2**
45 ÷ 5 = **9**	7 ÷ 7 = **1**	27 ÷ 9 = **3**	9 ÷ 1 = **9**	48 ÷ 6 = **8**	0 ÷ 7 = **0**
4 ÷ 1 = **4**	0 ÷ 9 = **0**	24 ÷ 3 = **8**	32 ÷ 4 = **8**	5 ÷ 5 = **1**	72 ÷ 9 = **8**
56 ÷ 7 = **8**	15 ÷ 3 = **5**	12 ÷ 6 = **2**	8 ÷ 2 = **4**	63 ÷ 7 = **9**	0 ÷ 4 = **0**
14 ÷ 2 = **7**	42 ÷ 6 = **7**	6 ÷ 1 = **6**	16 ÷ 8 = **2**	20 ÷ 5 = **4**	49 ÷ 7 = **7**
36 ÷ 4 = **9**	64 ÷ 8 = **8**	0 ÷ 3 = **0**	54 ÷ 9 = **6**	4 ÷ 2 = **2**	48 ÷ 8 = **6**
18 ÷ 9 = **2**	3 ÷ 1 = **3**	35 ÷ 5 = **7**	8 ÷ 4 = **2**	72 ÷ 8 = **9**	6 ÷ 6 = **1**
0 ÷ 5 = **0**	42 ÷ 7 = **6**	2 ÷ 2 = **1**	36 ÷ 9 = **4**	7 ÷ 1 = **7**	12 ÷ 3 = **4**

S

48 Divisions with Remainders

Divide.

3 R 3 4) 15	1 R 5 9) 14	6 R 3 7) 45	5 R 1 3) 16	6 R 2 6) 38	3 R 1 2) 7
6 R 2 8) 50	5 R 3 5) 28	5 R 1 4) 21	2 R 3 6) 15	1 R 4 7) 11	2 R 4 8) 20
6 R 2 3) 20	4 R 4 7) 32	3 R 6 8) 30	7 R 1 2) 15	8 R 3 5) 43	5 R 5 6) 35
6 R 8 9) 62	2 R 2 4) 10	4 R 3 6) 27	2 R 3 9) 21	4 R 3 4) 19	8 R 1 3) 25
9 R 2 6) 56	8 R 1 2) 17	3 R 1 3) 10	1 R 3 5) 8	4 R 4 9) 40	4 R 2 7) 30
2 R 1 2) 5	3 R 1 8) 25	3 R 2 5) 17	2 R 3 7) 17	2 R 2 3) 8	2 R 1 4) 9
2 R 6 7) 20	1 R 4 6) 10	4 R 1 2) 9	7 R 2 4) 30	1 R 7 8) 15	3 R 2 9) 29
6 R 2 5) 32	4 R 2 3) 14	5 R 5 9) 50	8 R 1 8) 65	5 R 1 2) 11	3 R 4 5) 19

T

60 Improper Fractions to Simplify

Write each improper fraction as a mixed number or whole number.

$\frac{15}{2} = \mathbf{7\frac{1}{2}}$	$\frac{9}{8} = \mathbf{1\frac{1}{8}}$	$\frac{10}{2} = \mathbf{5}$	$\frac{18}{6} = \mathbf{3}$	$\frac{8}{3} = \mathbf{2\frac{2}{3}}$	$\frac{12}{4} = \mathbf{3}$
$\frac{10}{10} = \mathbf{1}$	$\frac{3}{2} = \mathbf{1\frac{1}{2}}$	$\frac{11}{4} = \mathbf{2\frac{3}{4}}$	$\frac{4}{3} = \mathbf{1\frac{1}{3}}$	$\frac{12}{5} = \mathbf{2\frac{2}{5}}$	$\frac{5}{4} = \mathbf{1\frac{1}{4}}$
$\frac{12}{6} = \mathbf{2}$	$\frac{9}{3} = \mathbf{3}$	$\frac{5}{5} = \mathbf{1}$	$\frac{15}{4} = \mathbf{3\frac{3}{4}}$	$\frac{6}{2} = \mathbf{3}$	$\frac{9}{9} = \mathbf{1}$
$\frac{3}{3} = \mathbf{1}$	$\frac{7}{4} = \mathbf{1\frac{3}{4}}$	$\frac{21}{10} = \mathbf{2\frac{1}{10}}$	$\frac{11}{2} = \mathbf{5\frac{1}{2}}$	$\frac{7}{6} = \mathbf{1\frac{1}{6}}$	$\frac{24}{8} = \mathbf{3}$
$\frac{11}{3} = \mathbf{3\frac{2}{3}}$	$\frac{9}{5} = \mathbf{1\frac{4}{5}}$	$\frac{4}{2} = \mathbf{2}$	$\frac{21}{8} = \mathbf{2\frac{5}{8}}$	$\frac{6}{5} = \mathbf{1\frac{1}{5}}$	$\frac{12}{3} = \mathbf{4}$
$\frac{7}{2} = \mathbf{3\frac{1}{2}}$	$\frac{25}{6} = \mathbf{4\frac{1}{6}}$	$\frac{10}{9} = \mathbf{1\frac{1}{9}}$	$\frac{4}{4} = \mathbf{1}$	$\frac{12}{2} = \mathbf{6}$	$\frac{16}{15} = \mathbf{1\frac{1}{15}}$
$\frac{10}{5} = \mathbf{2}$	$\frac{5}{2} = \mathbf{2\frac{1}{2}}$	$\frac{7}{3} = \mathbf{2\frac{1}{3}}$	$\frac{8}{4} = \mathbf{2}$	$\frac{8}{8} = \mathbf{1}$	$\frac{27}{10} = \mathbf{2\frac{7}{10}}$
$\frac{16}{4} = \mathbf{4}$	$\frac{6}{6} = \mathbf{1}$	$\frac{25}{12} = \mathbf{2\frac{1}{12}}$	$\frac{5}{3} = \mathbf{1\frac{2}{3}}$	$\frac{7}{5} = \mathbf{1\frac{2}{5}}$	$\frac{16}{9} = \mathbf{1\frac{7}{9}}$
$\frac{15}{8} = \mathbf{1\frac{7}{8}}$	$\frac{10}{3} = \mathbf{3\frac{1}{3}}$	$\frac{33}{10} = \mathbf{3\frac{3}{10}}$	$\frac{2}{2} = \mathbf{1}$	$\frac{35}{6} = \mathbf{5\frac{5}{6}}$	$\frac{25}{8} = \mathbf{3\frac{1}{8}}$
$\frac{6}{3} = \mathbf{2}$	$\frac{8}{5} = \mathbf{1\frac{3}{5}}$	$\frac{9}{4} = \mathbf{2\frac{1}{4}}$	$\frac{12}{12} = \mathbf{1}$	$\frac{25}{2} = \mathbf{12\frac{1}{2}}$	$\frac{9}{2} = \mathbf{4\frac{1}{2}}$

U

40 Fractions to Reduce

Reduce each fraction to lowest terms.

$\frac{2}{10} = \mathbf{\frac{1}{5}}$	$\frac{8}{16} = \mathbf{\frac{1}{2}}$	$\frac{2}{6} = \mathbf{\frac{1}{3}}$	$\frac{10}{100} = \mathbf{\frac{1}{10}}$	$\frac{6}{8} = \mathbf{\frac{3}{4}}$
$\frac{10}{15} = \mathbf{\frac{2}{3}}$	$\frac{5}{10} = \mathbf{\frac{1}{2}}$	$\frac{8}{12} = \mathbf{\frac{2}{3}}$	$\frac{9}{15} = \mathbf{\frac{3}{5}}$	$\frac{4}{16} = \mathbf{\frac{1}{4}}$
$\frac{2}{8} = \mathbf{\frac{1}{4}}$	$\frac{4}{10} = \mathbf{\frac{2}{5}}$	$\frac{15}{20} = \mathbf{\frac{3}{4}}$	$\frac{4}{8} = \mathbf{\frac{1}{2}}$	$\frac{4}{6} = \mathbf{\frac{2}{3}}$
$\frac{6}{15} = \mathbf{\frac{2}{5}}$	$\frac{4}{12} = \mathbf{\frac{1}{3}}$	$\frac{25}{100} = \mathbf{\frac{1}{4}}$	$\frac{10}{25} = \mathbf{\frac{2}{5}}$	$\frac{12}{20} = \mathbf{\frac{3}{5}}$
$\frac{20}{100} = \mathbf{\frac{1}{5}}$	$\frac{6}{9} = \mathbf{\frac{2}{3}}$	$\frac{2}{4} = \mathbf{\frac{1}{2}}$	$\frac{3}{12} = \mathbf{\frac{1}{4}}$	$\frac{3}{15} = \mathbf{\frac{1}{5}}$
$\frac{3}{9} = \mathbf{\frac{1}{3}}$	$\frac{2}{12} = \mathbf{\frac{1}{6}}$	$\frac{6}{10} = \mathbf{\frac{3}{5}}$	$\frac{12}{16} = \mathbf{\frac{3}{4}}$	$\frac{50}{100} = \mathbf{\frac{1}{2}}$
$\frac{9}{12} = \mathbf{\frac{3}{4}}$	$\frac{3}{6} = \mathbf{\frac{1}{2}}$	$\frac{5}{15} = \mathbf{\frac{1}{3}}$	$\frac{10}{12} = \mathbf{\frac{5}{6}}$	$\frac{8}{24} = \mathbf{\frac{1}{3}}$
$\frac{12}{15} = \mathbf{\frac{4}{5}}$	$\frac{8}{10} = \mathbf{\frac{4}{5}}$	$\frac{75}{100} = \mathbf{\frac{3}{4}}$	$\frac{6}{12} = \mathbf{\frac{1}{2}}$	$\frac{12}{24} = \mathbf{\frac{1}{2}}$

V

50 Fractions to Simplify

Simplify these fractions. Reduce the answers.

$\frac{16}{20} = \frac{4}{5}$	$\frac{6}{4} = 1\frac{1}{2}$	$\frac{4}{6} = \frac{2}{3}$	$\frac{10}{8} = 1\frac{1}{4}$	$\frac{3}{12} = \frac{1}{4}$
$\frac{12}{9} = 1\frac{1}{3}$	$\frac{2}{4} = \frac{1}{2}$	$\frac{12}{10} = 1\frac{1}{5}$	$\frac{12}{4} = 3$	$\frac{12}{8} = 1\frac{1}{2}$
$\frac{8}{3} = 2\frac{2}{3}$	$\frac{8}{6} = 1\frac{1}{3}$	$\frac{4}{12} = \frac{1}{3}$	$\frac{10}{4} = 2\frac{1}{2}$	$\frac{4}{10} = \frac{2}{5}$
$\frac{20}{8} = 2\frac{1}{2}$	$\frac{4}{8} = \frac{1}{2}$	$\frac{20}{9} = 2\frac{2}{9}$	$\frac{24}{6} = 4$	$\frac{9}{6} = 1\frac{1}{2}$
$\frac{15}{10} = 1\frac{1}{2}$	$\frac{5}{2} = 2\frac{1}{2}$	$\frac{12}{20} = \frac{3}{5}$	$\frac{15}{9} = 1\frac{2}{3}$	$\frac{8}{12} = \frac{2}{3}$
$\frac{4}{20} = \frac{1}{5}$	$\frac{8}{24} = \frac{1}{3}$	$\frac{10}{6} = 1\frac{2}{3}$	$\frac{3}{6} = \frac{1}{2}$	$\frac{16}{10} = 1\frac{3}{5}$
$\frac{2}{8} = \frac{1}{4}$	$\frac{20}{6} = 3\frac{1}{3}$	$\frac{6}{3} = 2$	$\frac{25}{12} = 2\frac{1}{12}$	$\frac{9}{12} = \frac{3}{4}$
$\frac{10}{2} = 5$	$\frac{8}{8} = 1$	$\frac{50}{100} = \frac{1}{2}$	$\frac{6}{12} = \frac{1}{2}$	$\frac{15}{6} = 2\frac{1}{2}$
$\frac{10}{3} = 3\frac{1}{3}$	$\frac{10}{20} = \frac{1}{2}$	$\frac{24}{9} = 2\frac{2}{3}$	$\frac{6}{8} = \frac{3}{4}$	$\frac{16}{5} = 3\frac{1}{5}$
$\frac{5}{10} = \frac{1}{2}$	$\frac{14}{8} = 1\frac{3}{4}$	$\frac{15}{2} = 7\frac{1}{2}$	$\frac{21}{6} = 3\frac{1}{2}$	$\frac{16}{24} = \frac{2}{3}$

W 30 Percents to Write as Fractions

Write each percent as a reduced fraction.

$1\% = \frac{1}{100}$	$20\% = \frac{1}{5}$	$55\% = \frac{11}{20}$	$90\% = \frac{9}{10}$	$75\% = \frac{3}{4}$
$99\% = \frac{99}{100}$	$5\% = \frac{1}{20}$	$95\% = \frac{19}{20}$	$80\% = \frac{4}{5}$	$12\% = \frac{3}{25}$
$70\% = \frac{7}{10}$	$65\% = \frac{13}{20}$	$50\% = \frac{1}{2}$	$2\% = \frac{1}{50}$	$48\% = \frac{12}{25}$
$24\% = \frac{6}{25}$	$25\% = \frac{1}{4}$	$98\% = \frac{49}{50}$	$40\% = \frac{2}{5}$	$15\% = \frac{3}{20}$
$60\% = \frac{3}{5}$	$30\% = \frac{3}{10}$	$4\% = \frac{1}{25}$	$35\% = \frac{7}{20}$	$36\% = \frac{9}{25}$
$45\% = \frac{9}{20}$	$8\% = \frac{2}{25}$	$10\% = \frac{1}{10}$	$21\% = \frac{21}{100}$	$85\% = \frac{17}{20}$

X

24 Fraction-Decimal-Percent Equivalents

Write each fraction as a decimal and a percent.

Fraction	Decimal	Percent
$\frac{1}{100}$	**0.01**	**1%**
$\frac{1}{2}$	**0.5**	**50%**
$\frac{1}{20}$	**0.05**	**5%**
$\frac{1}{4}$	**0.25**	**25%**
$\frac{3}{4}$	**0.75**	**75%**
$\frac{1}{5}$	**0.2**	**20%**
$\frac{1}{50}$	**0.02**	**2%**
$\frac{1}{10}$	**0.1**	**10%**
$\frac{3}{10}$	**0.3**	**30%**
$\frac{2}{5}$	**0.4**	**40%**
$\frac{7}{10}$	**0.7**	**70%**
$\frac{3}{5}$	**0.6**	**60%**

Fraction	Decimal	Percent
$\frac{4}{5}$	**0.8**	**80%**
$\frac{9}{10}$	**0.9**	**90%**
$\frac{1}{3}$	Rounds to 0.333	**$33\frac{1}{3}$%**
$\frac{2}{3}$	Rounds to 0.667	**$66\frac{2}{3}$%**
$\frac{1}{8}$	**0.125**	**$12\frac{1}{2}$%**
$\frac{7}{8}$	**0.875**	**$87\frac{1}{2}$%**
$\frac{1}{25}$	**0.04**	**4%**
$\frac{1}{6}$	Rounds to 0.167	**$16\frac{2}{3}$%**
$\frac{1}{9}$	Rounds to 0.111	**$11\frac{1}{9}$%**
$\frac{3}{8}$	**0.375**	**$37\frac{1}{2}$%**
$\frac{5}{8}$	**0.625**	**$62\frac{1}{2}$%**
$\frac{5}{6}$	Rounds to 0.833	**$83\frac{1}{3}$%**

Pretest and Posttest

1. C. 43

2. B. 9.67

3. D. 22.25

4. A. 45,600

5. D. 36¢

6. B. 6

7. A. $2\frac{1}{2}$

8. C. 6

9. B. $1\frac{1}{2}$

10. C. $\frac{2}{3}$

11. D. 6

12. A. 2,500,000

13. C. 20 mm

14. B. 1 meter

15. D. 0.2

16. D. 144

17. C. 10

18. A. 10.3

19. B. 280,000

20. A. 0.01, 0.1, 1.0

21. C. $\frac{3}{9}$

22. D. 45 miles

23. B. 23 students

24. A. 80

25. B. $\frac{1}{2}$

26. C. $6.36

27. A. 128 ounces

28. B. $13.75

29. D. $\frac{4}{5}$

30. A. 300 miles

31. D. 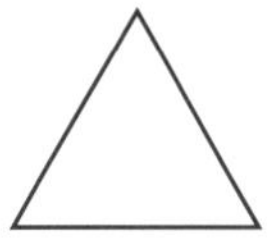

32. C. 

33. B. 70 mm

34. B. 20 inches

35. C. 96 sq. in.

36. A. $\frac{2}{5}$

37. C. 6

38. D. 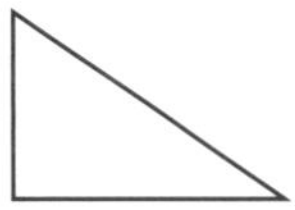

39. A. 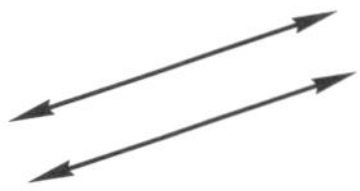

40. C. 9 sq. ft

41. B. $-2 < -1$

42. A. 27

43. D. 3

44. D. 5

45. C. $(6 \times 3) + (6 \times 4)$

46. C. 12

47. A. $\frac{1}{2} \times \frac{1}{2} < \frac{1}{2} + \frac{1}{2}$

48. B. 8

49. C. -3

50. D. Point *D*

Test 1, Form A

1. $6 + 12 = 18$ $\quad 12 + 6 = 18$
$18 - 12 = 6$ $\quad 18 - 6 = 12$

2. $15 \times 5 = 75$ $\quad 5 \times 15 = 75$
$75 \div 5 = 15$ $\quad 75 \div 15 = 5$

3. 5268

4. $2.15

5. $7.00

6. 40 quarters

7. 8 groups

8. 35 pencils

9. $16.45

10. 168

11. 3444

12. 27,024

13. 27,336

14. 1202

15. 125

16. 15

17. 12

18. 53

19. 742

20. 17

Test 1, Form B

1. $5 + 11 = 16$ $\quad 11 + 5 = 16$
   $16 - 11 = 5$ $\quad 16 - 5 = 11$

2. $4 \times 16 = 64$ $\quad 16 \times 4 = 64$
   $64 \div 4 = 16$ $\quad 64 \div 16 = 4$

3. $12.23

4. $6.51

5. $3.48

6. 15 nickels

7. 12 players

8. 31 days

9. 7402

10. 1460

11. 3839

12. 9200

13. 12,240

14. 1036

15. 144

16. 21

17. 4

18. 47

19. 1638

20. 18

Test 2, Form A

1. $2.76

2. $3.50

3. 5693

4. 2969

5. 1692

6. 28

7. $8.77

8. 18

9. 506

10. 63

11. 34

12. $\frac{3}{8}$

13. 40

14. $2\frac{3}{4}$ in.

15. >

16. 60 mm

17. 18

18. C. 4321

19. $4\overline{)20}$ $\quad 20 \div 4$ $\quad \frac{20}{4}$

20. $10 + 20 = 30$ $\quad 20 + 10 = 30$
    $30 - 20 = 10$ $\quad 30 - 10 = 20$

Test 2, Form B

1. $2.63

2. 3 cookies

3. 2112

4. 417

5. 1224

6. 602

7. $18.84

8. 3

9. 32

10. 35

11. 51

12. $\frac{3}{10}$

13. 18

14. 27 mm

15. =

16. 40 mm

17. 32

18. C. 7654

19. $8\overline{)16}$ $16 \div 8$ $\frac{16}{8}$

20. $11 + 22 = 33$ $22 + 11 = 33$
$33 - 22 = 11$ $33 - 11 = 22$

Test 3, Form A

1. 20 rolls

2. 561 years

3. ten thousands

4. ten million

5. 3

6. 431,720

7. 165 steps

8. 12°F

9. $\frac{2}{5}$

10. <

11. 42,650

12. 3216

13. $10.00

14. 506

15. 25

16. 8

17. 592

18. 7

19. 80 mm

20. $8 \times 15 = 120$ $15 \times 8 = 120$
$120 \div 8 = 15$ $120 \div 15 = 8$

Test 3, Form B

1. 13 players per team

2. 156 years

3. thousands

4. four billion

5. 11

6. 20

7. 176 steps

8. 2°F

9. $\frac{1}{8}$

10. >

11. $9.23

12. 4648

13. $3.50

14. 402

15. $4.17

16. 12

17. 1904

18. 6

19. 6 cm

20. $8 + 15 = 23$ $15 + 8 = 23$
$23 - 8 = 15$ $23 - 15 = 8$

Test 4, Form A

1. 2

2. 3

3. 216 pages
4. 9°
5. 65
6. 29
7. 38,000
8. 2500
9. 41
10. 8
11. 61
12. 1, 2, 3, 6, 9, 18
13. $9.75
14. 6
15. Friday
16. 5 answers
17. $1.75
18. 60 mm
19. 75
20. $2\frac{1}{2}$ in.

Test 4, Form B

1. 5
2. 5
3. 117 pages
4. 18°
5. 118
6. 26
7. 4400
8. 9000
9. 72
10. 8
11. 50
12. 1, 3, 7, 21
13. $9.25
14. 8
15. 22 points
16. 2 points
17. $3.75
18. 32 mm
19. 81
20. $1\frac{1}{2}$ in.

Test 5, Form A

1. 170
2. $\frac{13}{15}$
3. 17,105,000
4. 2700
5. B. 1632
6. $2\frac{2}{5}$
7. 60
8. 12
9. 6
10. $\frac{7}{12}$
11. 15
12. 8
13. $\frac{5}{6}$
14. $\frac{7}{12}$
15. $\frac{2}{5}$
16. 1
17. 50%
18. $\frac{3}{8}$
19. $2.75
20. >

Test 5, Form B

1. 300
2. $\frac{13}{16}$
3. 21,286,000
4. 3900
5. C. 3624
6. $5\frac{1}{3}$
7. 4610
8. 7
9. 5
10. $\frac{4}{15}$
11. 20
12. 18
13. $\frac{2}{3}$
14. $\frac{3}{8}$
15. $\frac{5}{8}$
16. 1
17. 25%
18. $\frac{7}{12}$
19. $3.75
20. <

Test 6, Form A

1. $6\frac{3}{4}$ inches
2. $\frac{7}{8}$
3. 1400
4. 63
5. 74
6. 1, 2, 4, 5, 10, 20
7. 8
8. B. diameter
9. 9 doughnuts
10. 0
11. $\frac{2}{3}$
12. $\frac{3}{8}$
13. $1\frac{2}{3}$
14. $6\frac{1}{3}$
15. $\frac{3}{4}$
16. 12
17. 9 in.
18. $\frac{3}{2}$
19. 40, 48, 56
20. $\angle C$ or $\angle ACB$ or $\angle BCA$

Test 6, Form B

1. $7\frac{1}{4}$ inches
2. $\frac{2}{3}$
3. 2000
4. 24
5. 123
6. 1, 3, 11, 33
7. 9
8. C. circumference
9. 15
10. $\frac{2}{5}$
11. $\frac{3}{4}$

12. $\frac{2}{15}$

13. $1\frac{1}{2}$

14. $8\frac{2}{5}$

15. $\frac{2}{3}$

16. 24

17. 15 cm

18. $\frac{4}{3}$

19. 49

20. $\angle B$ or $\angle ABC$ or $\angle CBA$

Test 7, Form A

1. 1800

2. $\frac{3}{2}$

3. 32 in.

4. 120 square tiles

5. 3

6. 0.012

7. 506

8. 12 in.

9. $\frac{47}{100}$

10. <

11. 0.35

12. $1\frac{2}{5}$

13. $\frac{2}{5}$

14. 30

15. 27

16. $\frac{1}{2}$

17. 110

18. $1\frac{1}{3}$

19. $\frac{4}{5}$

20. 32 answers

Test 7, Form B

1. 5300

2. $\frac{6}{5}$

3. 40 cm

4. 150 square tiles

5. 8

6. 1.5

7. 5070

8. 14 in.

9. $\frac{7}{10}$

10. >

11. 0.03

12. $1\frac{2}{7}$

13. $\frac{3}{5}$

14. 575

15. 30

16. $\frac{2}{3}$

17. 550

18. $2\frac{2}{3}$

19. $\frac{3}{5}$

20. 30 answers

Test 8, Form A

1. $\angle BDC$ or $\angle CDB$
2. 5
3. $\frac{4}{5}$
4. 9000
5. 19
6. 81
7. 4
8. $\frac{3}{5}$
9. $4\frac{1}{3}$
10. $\frac{3}{8}$
11. $\frac{1}{4}$
12. 1 hr 53 min
13. 24
14. 100 square-inch tiles
15. 3
16. 10.05
17. 1.5
18. 5.54
19. 0.09
20. 0.045

Test 8, Form B

1. $\angle ADB$ or $\angle BDA$
2. 2
3. $\frac{3}{4}$
4. 13,000
5. 29
6. 83
7. 7
8. $\frac{2}{5}$
9. 5
10. $3\frac{2}{3}$
11. $\frac{1}{3}$
12. 6 hr 15 min
13. 6
14. 48 square-inch tiles
15. 4
16. 3.12
17. 1.2
18. 10.56
19. 0.04
20. 0.048

Test 9, Form A

1. 45
2. $1\frac{1}{3}$
3. 9 in.
4. 180 sq. ft
5. 10.12
6. $20
7. 18
8. C. 70%
9. 4
10. $\frac{7}{100} = 0.07$
11. 4
12. $5\frac{1}{3}$
13. $\frac{1}{3}$

14. 15.13

15. 1.73

16. 0.0884

17. $1\frac{1}{3}$

18. 8

19. 0.175

20. C. $\overline{SR}$

Test 9, Form B

1. 60

2. $1\frac{1}{2}$

3. 25 cm

4. 132 sq. ft

5. 12.05

6. $15

7. 16

8. A. 30%

9. 7

10. $\frac{9}{100} = 0.09$

11. 24

12. $3\frac{3}{5}$

13. $\frac{1}{4}$

14. 13.68

15. 1.66

16. 0.0276

17. 7

18. 6

19. 0.375

20. D. $\overline{SQ}$

Test 10, Form A

1. 0.58

2. 3

3. $1.28

4. 8000

5. 2005

6. 96

7. $\frac{3}{4}$

8. 20

9. $\frac{5}{8}$

10. 12

11. 10,000 sq. cm

12. 31.40 in. or 31.4 in.

13. 10.2

14. 0.77

15. 10.12

16. 0.024

17. 0.041

18. 1.5

19. $4\frac{3}{5}$

20. 1250

Test 10, Form B

1. $\frac{7}{9}$

2. 7

3. $1.12

4. 1000

5. 209

6. 46

7. $\frac{2}{3}$

8. 36

9. $\frac{7}{8}$

10. 8

11. 1161 sq. mm

12. 31.40 ft or 31.4 ft

13. 21.05

14. 0.99

15. 17.07

16. 0.0156

17. 0.076

18. 4.1

19. $2\frac{2}{3}$

20. 350

Test 11, Form A

1. 25.12 cm

2. 0.0375

3. 88 mm

4. 480 sq. mm

5. $\frac{6}{5}$

6. 15 questions

7. 2.3

8. =

9. 80.08

10. 2.9

11. $5\frac{1}{3}$

12. $1\frac{2}{3}$

13. $\frac{1}{6}$

14. $\frac{2}{3}$

15. 0.0375

16. 0.0375

17. 10

18. $\frac{7}{8}$

19. $\frac{6}{7}$

20. 3.7

Test 11, Form B

1. 18.84 cm

2. 0.00972

3. 70 mm

4. 300 sq. mm

5. $\frac{4}{3}$

6. 10 questions

7. 2.9

8. >

9. 20.2

10. 6.73

11. 6

12. $3\frac{3}{5}$

13. $\frac{3}{10}$

14. $\frac{4}{5}$

15. 0.0045

16. 0.075

17. 5

18. $\frac{5}{8}$

19. $\frac{3}{10}$

20. 0.8

Test 12, Form A

1. $\frac{1}{3}$
2. 48 in.
3. 96 sq. cm
4. 3
5. 7.73
6. 0.71
7. 0.081
8. 94 inches
9. D. 9
10. $\frac{1}{2}$
11. $2\frac{2}{3}$
12. 0.05
13. 30
14. 123.4
15. $1\frac{1}{10}$
16. $\frac{11}{20}$
17. $>$
18. $6\frac{1}{6}$
19. $5\frac{1}{2}$
20. $0.66

Test 12, Form B

1. $\frac{1}{2}$
2. 42 in.
3. 108 sq. cm
4. 8
5. 18.76
6. 2.25
7. 0.084
8. 157 inches
9. C. 4
10. $\frac{2}{3}$
11. $\frac{3}{8}$
12. 0.24
13. 40
14. 1.234
15. $1\frac{1}{12}$
16. $\frac{7}{15}$
17. $<$
18. $7\frac{1}{4}$
19. $5\frac{2}{3}$
20. $0.76

Test 13, Form A

1. 4 sq. ft
2. $\frac{4}{4}$
3. $\frac{1}{3}$
4. $\frac{1}{2}$
5. D. 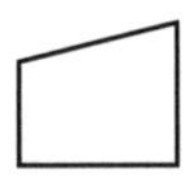
6. $2 \cdot 2 \cdot 2 \cdot 5$
7. $<$
8. $13.99
9. $5.94
10. 20.01
11. 2.66

12. 0.0486

13. 0.028

14. 15

15. $2\frac{1}{3}$

16. $3\frac{1}{2}$

17. $1\frac{5}{8}$

18. $3\frac{1}{3}$

19. 8 vertices

20. $\frac{5}{8}$

Test 13, Form B

1. 16 sq. in.

2. $\frac{3}{3}$

3. $\frac{2}{3}$

4. $\frac{1}{2}$

5. D.

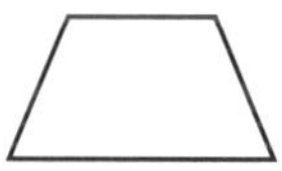

6. $2 \cdot 5 \cdot 5$

7. >

8. $15.89

9. $9.90

10. 33.67

11. 2.91

12. 0.21

13. 0.0375

14. 25

15. $3\frac{1}{2}$

16. $8\frac{1}{3}$

17. 1

18. $3\frac{1}{4}$

19. 12 edges

20. $\frac{2}{3}$

Test 14, Form A

1. 9

2. 121 sq. cm

3. 8.75

4. 0.0144

5. 55

6. 0.075

7. C.

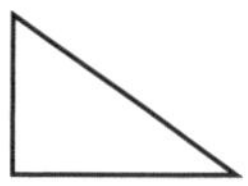

8. >

9. $26.45

10. 450

11. 15

12. $\frac{7}{10}$

13. $7\frac{1}{8}$

14. 8

15. 72 mm

16. $\frac{2 \cdot 2 \cdot 2 \cdot 3}{2 \cdot 3 \cdot 5} = \frac{4}{5}$

17. $3\frac{3}{4}$

18. $\frac{7}{8}$

19. C.

20. $\frac{1}{4}$

TEST 14, FORM B

1. 4
2. 144 sq. mm
3. 0.7
4. 0.0156
5. 45
6. 0.035
7. C. 
8. <
9. $20.49
10. 0.045
11. 7.5
12. $1\frac{5}{8}$
13. 14
14. 27
15. 6 cm
16. $\frac{2 \cdot 2 \cdot 2 \cdot 3}{2 \cdot 2 \cdot 3 \cdot 5} = \frac{2}{5}$
17. $2\frac{1}{3}$
18. $\frac{3}{5}$
19. B. 
20. $\frac{1}{3}$

TEST 15, FORM A

1. 24 sq. cm
2. $2.72
3. >
4. Point D
5. (3, –2)
6. B. pentagon
7. $\frac{3 \cdot 3 \cdot 5}{2 \cdot 3 \cdot 3 \cdot 3} = \frac{5}{6}$
8. 0.25
9. $0.12 = \frac{3}{25}$
10. 24
11. 1.82
12. 0.36
13. 12.5
14. 0.145
15. $1\frac{11}{12}$
16. $10\frac{1}{10}$
17. $2\frac{3}{8}$
18. 4
19. 3
20. 4.3

TEST 15, FORM B

1. 72 sq. cm
2. $3.63
3. <
4. Point H
5. (2, –3)
6. 8 sides
7. $\frac{2 \cdot 3 \cdot 3 \cdot 3}{3 \cdot 3 \cdot 3 \cdot 3} = \frac{2}{3}$
8. 0.8
9. $0.15 = \frac{3}{20}$
10. 24
11. 2.95
12. 0.09
13. 8

14. 0.125

15. $1\frac{5}{8}$

16. $11\frac{1}{3}$

17. $3\frac{11}{12}$

18. 12

19. 3

20. 10.8

Test 16, Form A

1. $\frac{1}{4}$

2. 84%

3. 24 sq. cm

4. 10 units

5. 6 sq. units

6. >

7. 8 eggs.

8. False

9. $\frac{3}{5}$, 18 girls

10. 10

11. 16.02

12. 0.35

13. 0.036

14. $\frac{1}{2}$

15. $7\frac{1}{4}$

16. $\frac{1}{2}$

17. 3

18. 36

19. $1\frac{1}{5}$

20. 0.57

Test 16, Form B

1. 2 quarts

2. 86%

3. 6 sq. cm

4. 12 units

5. 9 sq. units

6. <

7. 10 eggs

8. True

9. $\frac{2}{5}$, 24 minutes

10. 10

11. 23.17

12. 0.68

13. 0.052

14. $\frac{1}{2}$

15. $8\frac{1}{8}$

16. $3\frac{5}{8}$

17. 3

18. 28

19. $\frac{6}{7}$

20. 5.5

Test 17, Form A

1. $\frac{4}{5} = 0.80 = 0.8$

2. 48 cm^2

3. 60%

4. 314 cm

5. 44 mm

6. 90 sq. mm

7. 50°

8. $\frac{3}{2}$

9. (–1, 2)

10. 1500 tickets

11. 26 in.

12. 22

13. 360 in.^{3}

14. 15

15. 8.9

16. 0.15

17. $5\frac{3}{8}$

18. $1\frac{3}{4}$

19. $4\frac{1}{2}$

20. $1\frac{1}{3}$

Test 17, Form B

1. $\frac{3}{5}$ = 0.60 = 0.6

2. 48 cm^2

3. 80%

4. 628 cm

5. 36 cm

6. 60 sq. cm

7. 50°

8. $\frac{4}{3}$

9. (–2, 3)

10. 16 questions

11. 30 in.

12. 2

13. 200 in.^{3}

14. 15

15. 1.78

16. 0.14

17. $3\frac{1}{4}$

18. $2\frac{3}{4}$

19. 8

20. $\frac{3}{4}$

Test 18, Form A

1. 50 seeds

2. 8 girls

3. 24 cm

4. 24 sq. cm

5. 1.79

6. 0.081

7. 0.045

8. 30

9. C. 4 and 5

10. $1\frac{1}{3}$

11. $\frac{5}{6}$

12. 6

13. 4

14. $1\frac{4}{5}$

15. 62.8 cm

16. 314 cm^2

17. $\frac{3}{4}$ = 0.75

18. 300 cm^3

19. 6 faces

20. (–2, –2), (8, –2), or (2, 6)

Test 18, Form B

1. 40 seeds
2. 18 boys
3. 12 cm
4. 6 sq. cm
5. 1.7
6. 0.086
7. 0.075
8. 50
9. C. 5 and 6
10. $1\frac{1}{2}$
11. $\frac{3}{4}$
12. 2
13. 3
14. $1\frac{3}{5}$
15. 62.8 in.
16. 314 in.^{2}
17. $\frac{13}{20} = 0.65$
18. 1000 cm^3
19. 12 edges
20. (1, 0), (–5, 0), or (5, 6)

Test 19, Form A

1. $\frac{1}{2}$
2. 90
3. 25%
4. 8 edges
5. 54 sq. cm
6. C. right triangle
7. 26
8. $2\frac{1}{4}$
9. 15
10. 12
11. 12
12. 35
13. 4.43
14. 8
15. $14\frac{1}{6}$
16. $\frac{7}{12}$
17. $\frac{3}{5}$
18. $33\frac{1}{3}\%$
19. 3.75; 5.25
20. 20 units

Test 19, Form B

1. $\frac{1}{2}$
2. 19
3. 40%
4. 5 faces
5. 30 sq. cm
6. D. right triangle
7. 7
8. $6\frac{1}{4}$
9. 12
10. 24
11. 30
12. 36
13. 6.83
14. 6

15. $10\frac{1}{8}$

16. $\frac{11}{12}$

17. $1\frac{1}{3}$

18. $66\frac{2}{3}\%$

19. 3.75; 6.25

20. 18 units

Test 20, Form A

1. 16 B's
2. $\frac{1}{36}$
3. (a) $\frac{3}{5}$ (b) 60%
4. (a) $\frac{1}{20}$ (b) 0.05
5. 12 students
6. 180 sq. mm
7. 56 mm
8. 80°
9. 360°
10. 30
11. –8
12. 2
13. 60 cm^3
14. 4
15. 11
16. 1.75
17. 15
18. 1.4
19. 18°F
20. 113.04 $in.^2$

Test 20, Form B

1. 9 B's
2. $\frac{1}{36}$
3. (a) $\frac{2}{5}$ (b) 40%
4. (a) $\frac{1}{25}$ (b) 0.04
5. 10 students
6. 48 sq. cm
7. 32 cm
8. 105°
9. 360°
10. 20
11. –7
12. 4
13. 27 cm^3
14. 2
15. 5
16. 10.75
17. 12
18. 2.3
19. 13°F
20. 200.96 $in.^2$

Test 21, Form A

1. $0.16
2. 25 students
3. (1, –2)
4. 61
5. 24 sugar cubes
6. 24 cm
7. 62.8 cm
8. 4000 pounds

9. 15

10. 15 boys

11. (a) 1.6 (b) 160%

12. (a) $\frac{9}{10}$ (b) 90%

13. (a) $\frac{1}{10}$ (b) 0.1

14. 3.05

15. –1

16. 10

17. $10\frac{5}{12}$

18. 10

19. $1\frac{1}{2}$

20. 24 sq. cm

Test 21, Form B

1. $0.81

2. 30 students

3. (2, 3)

4. 133

5. 27 sugar cubes

6. 34 cm

7. 314 cm^2

8. 1000 pounds

9. 18

10. 12 boys

11. (a) 2.2 (b) 220%

12. (a) $\frac{7}{10}$ (b) 70%

13. (a) $\frac{1}{20}$ (b) 0.05

14. 5.45

15. 0

16. 0.0024

17. $3\frac{5}{12}$

18. 15

19. $1\frac{1}{3}$

20. 30 sq. cm

Test 22, Form A

1. 40°

2. 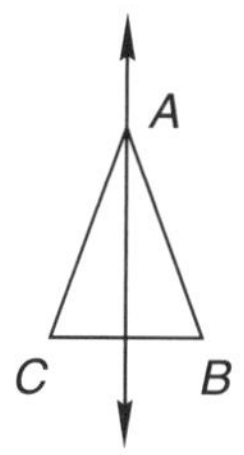

3. $2\frac{2}{3}$

4. $5\frac{1}{3}$

5. 4.125

6. –2, 0.2, $\frac{1}{2}$

7. 12 sq. units

8. –13

9. $\frac{1}{15}$

10. 13

11. (a) 1.2 (b) 120%

12. (a) $\frac{3}{10}$ (b) 30%

13. (a) $\frac{7}{20}$ (b) 0.35

14. 80 cu. in.

15. 314 $in.^2$

16. $\frac{7}{9}$

17. 8

18. 11

19. 36 m

20. 50 sq. m

Test 22, Form B

1. 30°

2. 

3. $1\frac{5}{6}$

4. 3

5. 4.625

6. $-3, 0.35, \frac{3}{4}$

7. 12 sq. units

8. -12

9. $\frac{2}{9}$

10. 7

11. (a) 1.25 (b) 125%

12. (a) $\frac{7}{10}$ (b) 70%

13. (a) $\frac{3}{20}$ (b) 0.15

14. 96 blocks

15. 100.48 in.

16. $\frac{9}{7}$

17. 12

18. 9

19. 50 m

20. 118 sq. m

Test 23, Form A

1. 27, 27, 28, 28

2. 1,500,000

3. 3^4

4. $\overset{2}{\cancel{8}}\ \cancel{\text{qt}} \times \frac{1\text{ gal}}{\underset{1}{\cancel{4}}\ \cancel{\text{qt}}} = \mathbf{2\ gal}$

5. 5040

6. $\frac{1}{5}$

7. 45 cm

8. 15 ft

9. 31.4 in.

10. 42 m

11. 80 sq. m

12. (a) 0.75 (b) 75%

13. (a) $\frac{1}{50}$ (b) 2%

14. (a) $1\frac{1}{5}$ (b) 1.2

15. 5.025

16. 3

17. 28

18. -8

19. 40 sq. cm

20. 25 questions

Test 23, Form B

1. 28, 28, 29

2. 1,500,000,000

3. 2^6

4. $8\ \cancel{\text{qt}} \times \frac{2\text{ pt}}{1\ \cancel{\text{qt}}} = \mathbf{16\ pt}$

5. 640

6. $\frac{1}{5}$

7. 12 in.

8. 25 sq. cm

9. 62.8 in.

10. 32 cm

11. 48 sq. cm

12. (a) 0.25 (b) 25%

13. (a) $\frac{1}{25}$ (b) 4%

14. (a) $1\frac{1}{2}$ (b) 1.5

15. 9.73

16. 30

17. 36

18. –6

19. 48 sq. mm

20. 30 questions

Final Exam

1. C. $304\frac{2}{5}$

2. D. 19.25

3. A. 1.55

4. A. 0.63

5. B. 5.2

6. B. $4\frac{1}{2}$

7. C. $1\frac{9}{10}$

8. C. $2\frac{2}{3}$

9. A. $1\frac{3}{5}$

10. C. $\frac{5}{7}$

11. B. 4

12. A. 2,500,000

13. C. 1200 mm

14. B. 2 m

15. D. 0.04

16. B. 10^7

17. C. 14 and 15

18. E. None correct

19. D. 26

20. C. 0.3, 33%, $\frac{1}{3}$

21. C. 18

22. D. 30

23. D. 31

24. B. 9

25. C. $\frac{1}{2}$

26. B. $3.12

27. C. 13¢/oz

28. C. $27.00

29. B. 84%

30. D. 1000 miles

31. B.

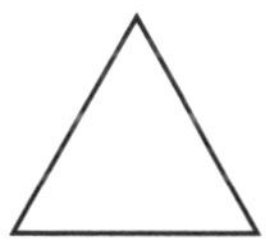

32. A. 52°

33. D. 70 cm

34. B. 62.8 in.

35. B. 54 m^2

36. A. 50 ft^2

37. A. 80 cm^3

38. C.

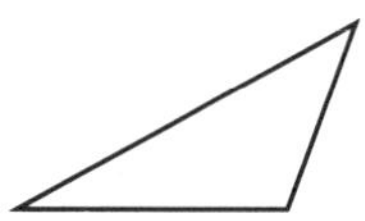

39. D. $\overline{TS}$

40. A. 5

41. B. –7

42. D. 144

43. C. 36

44. A. $2^3 \cdot 3$

45. D. $6 \cdot 20 + 6 \cdot 5$

46. B. 12

47. A. $\sqrt{9} < 2^2$

48. B. 100

49. D. –3

50. A. (–2, –3)

Test-Day Activity Master 2

1. (a) Geometric sequence
   (b) The next term is found by multiplying the previous term by 2.
   (c) 48, 96, 192
2. (a) Arithmetic sequence
   (b) The next term is found by adding 4 to the previous term.
   (c) 33, 37, 41
3. 76, 70
4. 22, 29
5. MN, PQ

Test-Day Activity Master 3

1. 35, 40, 45
2. 49, 64, 81
3. 45, 55, 66
4. F, S, S
5. 33, 44, 66
6. 4, 8, 12
7. 1, 2, 4, 8, 16, ... (Sample answer)
8. 1, 3, 5, 7 9, ... (Sample answer)
9. You can find the next term by adding 7 to the previous term in the sequence. (Sample answer)
10. 21, 34. You can find the next term by adding the two previous terms in the sequence. (Sample answer)

Test-Day Activity Master 4

1. $12 - 3 + 4 = 13$ (Sample answer)
2. $1 \times 3 \times 5 = 15$ (Sample answer)
3. $9 \times 10 \div 9 = 10$ (Sample answer)
4. $5 + (3 \times 4) = 17$ (Sample answer)
5. $10 + 11 - 12 = 9$ (Sample answer)
6. $24 \div (6 \times 2) = 2$ (Sample answer)
7. $(6 - 0) \times 7 = 42$ (Sample answer)

Test-Day Activity Master 5

1. Positive number
2. Negative number
3. Negative number
4. Negative number
5. Positive number
6. Negative number

Test-Day Activity Master 6

1. –50 dollars
2. –3 seconds
3. 20,320 feet
4. –4 strokes
5. 16 pounds
6. –150 feet
7. The temperature outside is 10° below zero. (Sample answer)
8. The football team lost 14 yards on the play. (Sample answer)
9. Sam borrowed $200. (Sample answer)
10. The desert floor was 164 feet below sea level. (Sample answer)

Test-Day Activity Master 8

e. 54 sq. in. This is the area of the chosen pattern. The outside of the cube is made from the pattern so the surface area of the cube and the area of the pattern are the same. (Sample answer)

Test-Day Activity Master 11

1. 36.2 cm
2. 28.09 sq. cm
3. A and B or C and E
4. D, C, and E or C, E, and F or C, G, and E
5. C, G, and E

Test-Day Activity Master 12

52 sq. in. This is the area of the chosen pattern. The outside of the box is made from the pattern so the surface area of the box and the area of the pattern are the same. (Sample answer)

Test-Day Activity Master 17

1. Mean: 88; median: 86; mode: 86
2. The mean is the best measure central tendency for this situation since Jeff's scores are fairly consistent. There are no really high scores or really low scores.

Test-Day Activity Master 18

1. The median provides more useful information about the price of the typical home. Most of the homes are priced near the median. The presence of an outlier has a greater effect on the mean than on the median.
2. The mode provides the most useful information for helping the purchaser decide what equipment to order because it identifies the weights of the bowling balls most used by local bowlers. The equipment purchaser will want to order more 10- and 16-pound bowling balls than those of other weights.

Test-Day Activity Master 21

1. Exact; paper/pencil or calculator; $3.75
2. Exact; mental;
   $3\frac{1}{2}$ c; $\frac{1}{2}$ tsp; 4 tsp; 1 tsp; $\frac{2}{3}$ c; $1\frac{1}{3}$ c; 4 eggs; 2 c
3. Estimate; mental; about 15 miles

Test-Day Activity Master 22

Front	Top	Right Side
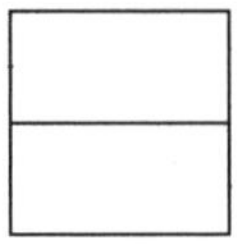	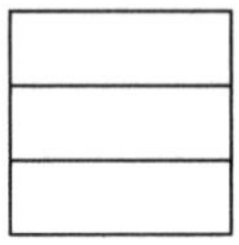	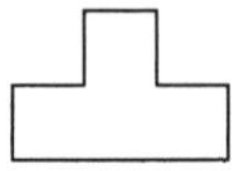

Test-Day Activity Master 23

Front	Top	Right Side
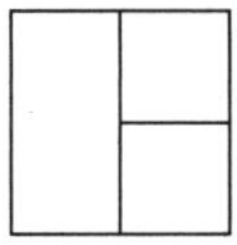	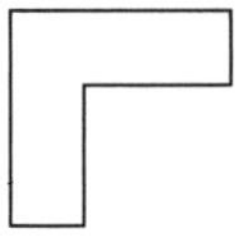	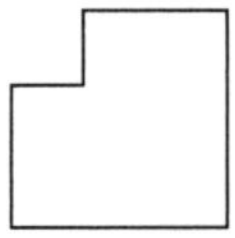

Test-Day Activity Master 25

2 cups

4 cups; 2 pints

16 cups; 8 pints; 4 quarts

Test-Day Activity Master 27

1. 3125; 5 × ====
2. 1331; 11 × ==
3. 1024; 2 × =========
4. 4096; 8 × ===
5. 729
6. 4,782,969
7. 262,144
8. 100

Test-Day Activity Master 30

1. 60°
2. 54°
3. 25°
4. 79°
5. 110°
6. 90°
7. 100°
8. 60°

Test-Day Activity Master 31

1. 26 in.
2. $P = 2l + 2w$
3. 42 cm^2
4. $A = l \cdot w$

Test-Day Activity Master 32

1. 36 m^2
2. $A = s^2$
3. 336 $ft.^3$
4. $V = lwh$

Test-Day Activity Master 33

1. Exact; paper and pencil calculation; a minimum of $16\frac{2}{3}$ yards
2. Exact; paper and pencil or mental calculation; he needs to buy more
3. Estimate; mental; about 2 hours total

Test-Day Activity Master 34

1. Hole 8
2. Hole 4
3. One stroke under par—3 strokes
4. Below par

Test-Day Activity Master 35

1. March 29
2. May 17
3. Yes. Over the 5 dates, Sam has deposited $575 and withdrawn $492.

Test-Day Activity Master 37

1. mg
2. kg
3. kg
4. g
5. g
6. kg
7. kg
8. g
9. mg
10. g

Activity Master 1

See student work.

Activity Master 2

~~1~~	2	3	~~4~~	5	~~6~~	7	~~8~~	~~9~~	~~10~~
11	~~12~~	13	~~14~~	~~15~~	~~16~~	17	~~18~~	19	~~20~~
~~21~~	~~22~~	23	~~24~~	~~25~~	~~26~~	~~27~~	~~28~~	29	~~30~~
31	~~32~~	~~33~~	~~34~~	~~35~~	~~36~~	37	~~38~~	~~39~~	~~40~~
41	~~42~~	43	~~44~~	~~45~~	~~46~~	47	~~48~~	~~49~~	~~50~~
~~51~~	~~52~~	53	~~54~~	~~55~~	~~56~~	~~57~~	~~58~~	59	~~60~~
61	~~62~~	~~63~~	~~64~~	~~65~~	~~66~~	67	~~68~~	~~69~~	~~70~~
71	~~72~~	73	~~74~~	~~75~~	~~76~~	~~77~~	~~78~~	79	~~80~~
~~81~~	~~82~~	83	~~84~~	~~85~~	~~86~~	~~87~~	~~88~~	89	~~90~~
~~91~~	~~92~~	~~93~~	~~94~~	~~95~~	~~96~~	97	~~98~~	~~99~~	~~100~~

Activity Master 8

1. 45°
2. 90°
3. 30°
4. 150°
5. 135°
6. 165°
7. 60°
8. 165°
9. 90°
10. 180°
11. 105°
12. 150°

Activity Master 9

See student work.

Activity Master 11

Section A

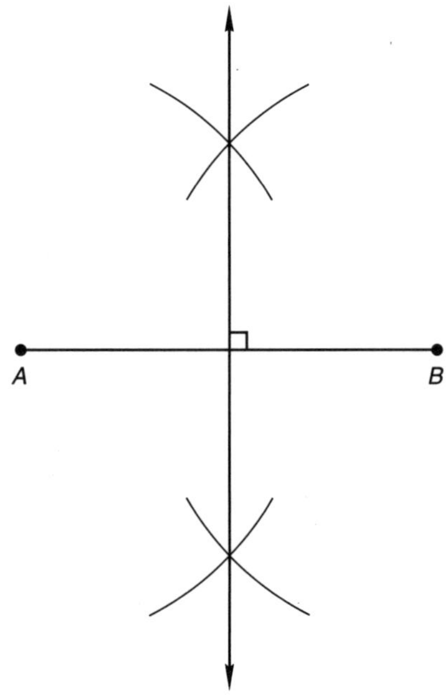

Section B

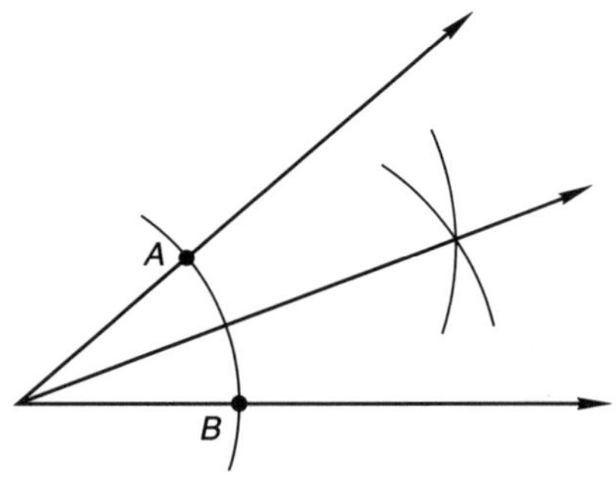

Pretest and Posttest

1. $$\begin{array}{r} 43 \\ 22\overline{)946} \\ \underline{88} \\ 66 \\ \underline{66} \\ 0 \end{array}$$

   C. 43

2. $$\begin{array}{r} \overset{1}{6}.50 \\ 2.47 \\ \underline{+\ 0.70} \\ 9.67 \end{array}$$

   B. 9.67

3. $$\begin{array}{r} 23.45 \\ \underline{-\ \ 1.20} \\ 22.25 \end{array}$$

   D. 22.25

4. $$\begin{array}{r} 608 \\ \underline{\times\ \ 75} \\ 3040 \\ \underline{4256\ } \\ 45{,}600 \end{array}$$

   A. 45,600

5. $$\begin{array}{r} \$0.36 \\ 20\overline{)\$7.20} \\ \underline{6\ 0} \\ 1\ 20 \\ \underline{1\ 20} \\ 0 \end{array}$$

   D. 36¢

6. $$\begin{array}{r} 3\frac{2}{3} \\ \underline{+\ 2\frac{1}{3}} \\ 5\frac{3}{3} \end{array} = 6$$

   B. 6

7. $$\begin{array}{r} 4\frac{3}{4} \\ \underline{-\ 2\frac{1}{4}} \\ 2\frac{2}{4} \end{array} = 2\frac{1}{2}$$

   A. $2\frac{1}{2}$

8. $\frac{3}{5} \times 10 = \frac{3}{5} \times \frac{10}{1} = \frac{30}{5} = 6$

   C. 6

9. $\frac{3}{4} \div \frac{1}{2} = \frac{3}{4} \cdot \frac{2}{1} = \frac{6}{4} = \frac{3}{2} = 1\frac{1}{2}$

   B. $1\frac{1}{2}$

10. $\frac{8}{12} = \frac{8 \div 4}{12 \div 4} = \frac{2}{3}$

    C. $\frac{2}{3}$

11. **D. 6**

12. **A. 2,500,000**

13. $2\ \cancel{\text{cm}} \cdot \frac{10\text{ mm}}{1\ \cancel{\text{cm}}} = 20\text{ mm}$

    C. 20 mm

14. **B. 1 meter**

15. **D. 0.2**

16. $12^2 = 12 \cdot 12 = 144$

    D. 144

17. $\sqrt{100} = \sqrt{10 \cdot 10} = 10$

    C. 10

18. **A. 10.3**

19. $692 \approx 700$
    $412 \approx 400$
    $700 \cdot 400 = 280{,}000$

    B. 280,000

20. $0.01 < 0.1 < 1.0$

    A. 0.01, 0.1, 1.0

21. $\frac{1}{3} \cdot \frac{3}{3} = \frac{3}{9}$

    C. $\frac{3}{9}$

22.

Ratio	Actual Count
15 miles	*N*
1 hour	3 hours

$\frac{15}{1} = \frac{N}{3}$

$N = 45$ miles

D. 45 miles

23. $\frac{22 + 22 + 25}{3} = \frac{69}{3} = 23$

B. 23 students

24. 80 occurs most often, so it is the mode.

A. 80

25. **B.** $\frac{1}{2}$

26.

$$\begin{array}{r} \overset{1}{\$5}.\overset{1}{8}7 \\ +\ \$0.49 \\ \hline \$6.36 \end{array}$$

C. $6.36

27.

Ratio	Actual Count
1 pound	8 pounds
16 ounces	*N*

$\frac{1}{16} = \frac{8}{N}$

$N = 128$

A. 128 ounces

28.

$$\begin{array}{r} \$13.75 \\ 2\overline{)\$27.50} \\ \underline{2} \\ 07 \\ \underline{6} \\ 1\,5 \\ \underline{1\,4} \\ 10 \\ \underline{10} \\ 0 \end{array}$$

B. $13.75

29. $\frac{16}{20} \div \frac{4}{4} = \frac{4}{5}$

D. $\frac{4}{5}$

30.

Ratio	Actual Count
60 miles	*N*
1 hour	5 hours

$\frac{60}{1} = \frac{N}{5}$

$N = 300$

A. 300 miles

31. **D.**

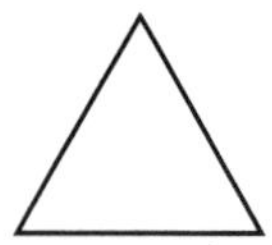

32. **C.**

33. Perimeter = 20 mm + 15 mm + 20 mm + 15 mm
= 70 mm

B. 70 mm

34. Radius = Diameter ÷ 2
= 40 in. ÷ 2
= 20 in.

B. 20 inches

35. Area = 12 in. × 8 in. = 96 in.^{2}

C. 96 sq. in.

36. **A.** $\frac{2}{5}$

37. **C. 6**

38. **D.**

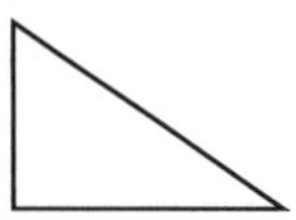

39. **A.**

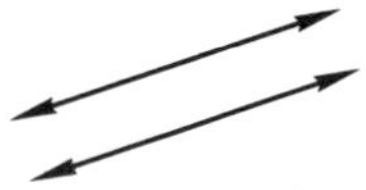

40. Area = 3 ft × 3 ft = 9 ft^2

C. 9 sq. ft

41. **B. –2 < –1**

42.

$$\begin{array}{r} 15 \\ +\ 12 \\ \hline 27 \end{array}$$

A. 27

43. $5\frac{2}{3} - \left(4 - 1\frac{1}{3}\right) = 5\frac{2}{3} - 2\frac{2}{3} = 3$

D. 3

44. **D. 5**

45. 6 × 12 = 72
(6 × 10) + (6 × 2) = 60 + 12 = 72
(6 × 6) + (6 × 6) = 36 + 36 = 72
(6 × 3) + (6 × 4) = 18 + 24 = 42
(3 × 12) + (3 × 12) = 36 + 36 = 72

C. (6 × 3) + (6 × 4)

46. 36 ÷ 3 = 12

C. 12

47. **A.** $\mathbf{\frac{1}{2} \times \frac{1}{2} < \frac{1}{2} + \frac{1}{2}}$

48. $x = \frac{12}{2} = 6$

$2 + x = 2 + 6 = 8$

B. 8

49. **C. –3**

50. **D. Point *D***

TEST 1, FORM A

1. **6 + 12 = 18 12 + 6 = 18**
18 – 12 = 6 18 – 6 = 12

2. **15 × 5 = 75 5 × 15 = 75**
75 ÷ 5 = 15 75 ÷ 15 = 5

3.

$$\begin{array}{r} 3675 \\ 285 \\ +\ 1308 \\ \hline \mathbf{5268} \end{array}$$

4.

$$\begin{array}{r} \$5.00 \\ -\ \$2.85 \\ \hline \mathbf{\$2.15} \end{array}$$

5.

$$\begin{array}{r} \$1.75 \\ \times\quad 4 \\ \hline \mathbf{\$7.00} \end{array}$$

6. 4 quarters × 10 = **40 quarters**

7. 240 ÷ 30 = **8 groups**

8. Three dozen (3 × 12) is 36.
36 – 1 = **35 pencils**

9.

$$\begin{array}{r} \$3.50 \\ \$0.95 \\ +\ \$12.00 \\ \hline \mathbf{\$16.45} \end{array}$$

10. 28 + 28 + 28 + 28 + 28 + 28

$$\begin{array}{r} 28 \\ \times\ \ 6 \\ \hline \mathbf{168} \end{array}$$

11.

$$\begin{array}{r} 3826 \\ -\ 382 \\ \hline \mathbf{3444} \end{array}$$

12.

$$\begin{array}{r} 563 \\ \times\ 48 \\ \hline 4504 \\ 2252 \\ \hline \mathbf{27{,}024} \end{array}$$

13.
$$\begin{array}{r} 408 \\ \times\ 67 \\ \hline 2856 \\ 2448 \\ \hline \mathbf{27{,}336} \end{array}$$

14.
$$\begin{array}{r} \mathbf{1202} \\ 8\overline{)9616} \\ \underline{8} \\ 16 \\ \underline{16} \\ 016 \\ \underline{16} \\ 0 \end{array}$$

15.
$$\begin{array}{r} \mathbf{125} \\ 24\overline{)3000} \\ \underline{24} \\ 60 \\ \underline{48} \\ 120 \\ \underline{120} \\ 0 \end{array}$$

16. $20 - (10 - 5) = 20 - 5 = \mathbf{15}$

17. $36 \div (6 \div 2) = 36 \div 3 = \mathbf{12}$

18. $47 + m = 100$
$$\begin{array}{r} 100 \\ -\ 47 \\ \hline \mathbf{53} \end{array}$$

19. $1000 - n = 258$
$$\begin{array}{r} 1000 \\ -\ 258 \\ \hline \mathbf{742} \end{array}$$

20. $5x = 85$

$x = \dfrac{85}{5}$
$$\begin{array}{r} \mathbf{17} \\ 5\overline{)85} \\ \underline{5} \\ 35 \\ \underline{35} \\ 0 \end{array}$$

Test 1, Form B

1. **5 + 11 = 16** **11 + 5 = 16**
16 − 11 = 5 **16 − 5 = 11**

2. **4 × 16 = 64** **16 × 4 = 64**
64 ÷ 4 = 16 **64 ÷ 16 = 4**

3.
$$\begin{array}{r} \$1.25 \\ \$0.98 \\ +\ \$10.00 \\ \hline \mathbf{\$12.23} \end{array}$$

4.
$$\begin{array}{r} \$10.00 \\ -\ \ \$3.49 \\ \hline \mathbf{\$6.51} \end{array}$$

5.
$$\begin{array}{r} \$0.29 \\ \times\ \ \ \ 12 \\ \hline 58 \\ 29 \\ \hline \mathbf{\$3.48} \end{array}$$

6. One quarter equals 5 nickels, so 3 quarters equals $3 \times 5 =$ **15 nickels.**

7. $60 \div 5 =$ **12 players**

8. $(4 \times 7) + 3 = 28 + 3 =$ **31 days**

9.
$$\begin{array}{r} 5986 \\ 367 \\ +\ 1049 \\ \hline \mathbf{7402} \end{array}$$

10. $365 + 365 + 365 + 365$
$$\begin{array}{r} 365 \\ \times\ \ \ 4 \\ \hline \mathbf{1460} \end{array}$$

11.
$$\begin{array}{r} 4216 \\ -\ \ 377 \\ \hline \mathbf{3839} \end{array}$$

12.
$$\begin{array}{r} 575 \\ \times\ \ 16 \\ \hline 3450 \\ 575 \\ \hline \mathbf{9200} \end{array}$$

13.
$$\begin{array}{r} 306 \\ \times\ \ 40 \\ \hline \mathbf{12{,}240} \end{array}$$

14.
$$\begin{array}{r} \mathbf{1036} \\ 7\overline{)7252} \\ \underline{7} \\ 025 \\ \underline{21} \\ 42 \\ \underline{42} \\ 0 \end{array}$$

15. $$\begin{array}{r} \mathbf{144} \\ 25\overline{)3600} \\ \underline{25} \\ 110 \\ \underline{100} \\ 100 \\ \underline{100} \\ 0 \end{array}$$

16. $56 - (23 + 12) = 56 - 35 = \mathbf{21}$

17. $48 \div (3 \times 4) = 48 \div 12 = \mathbf{4}$

18. $67 + m = 114$ $$\begin{array}{r} 114 \\ \underline{-\ \ 67} \\ \mathbf{47} \end{array}$$

19. $3000 - n = 1362$ $$\begin{array}{r} 3000 \\ \underline{-\ 1362} \\ \mathbf{1638} \end{array}$$

20. $4x = 72$
$x = \dfrac{72}{4}$ $$\begin{array}{r} \mathbf{18} \\ 4\overline{)72} \\ \underline{4} \\ 32 \\ \underline{32} \\ 0 \end{array}$$

Test 2, Form A

1. $$\begin{array}{r} \$0.49 \\ \$1.29 \\ \underline{+\ \$0.98} \\ \mathbf{\$2.76} \end{array}$$

2. $\$0.35 \times 10 = \mathbf{\$3.50}$

3. $$\begin{array}{r} 3496 \\ 788 \\ \underline{+\ 1409} \\ \mathbf{5693} \end{array}$$

4. $$\begin{array}{r} 4806 \\ \underline{-\ 1837} \\ \mathbf{2969} \end{array}$$

5. $$\begin{array}{r} 47 \\ \underline{\times\ 36} \\ 282 \\ \underline{141} \\ \mathbf{1692} \end{array}$$

6. $$\begin{array}{r} \mathbf{28} \\ 15\overline{)420} \\ \underline{30} \\ 120 \\ \underline{120} \\ 0 \end{array}$$

7. $$\begin{array}{r} \$5.00 \\ \$3.49 \\ \underline{+\ \$0.28} \\ \mathbf{\$8.77} \end{array}$$

8. $24 - (12 - 6) = 24 - 6 = \mathbf{18}$

9. $$\begin{array}{r} \mathbf{506} \\ 8\overline{)4048} \\ \underline{40} \\ 048 \\ \underline{48} \\ 0 \end{array}$$

10. $67 + n = 130$ $$\begin{array}{r} 130 \\ \underline{-\ \ 67} \\ \mathbf{63} \end{array}$$

11. $9m = 306$
$m = \dfrac{306}{9}$ $$\begin{array}{r} \mathbf{34} \\ 9\overline{)306} \\ \underline{27} \\ 36 \\ \underline{36} \\ 0 \end{array}$$

12. Three out of eight is $\mathbf{\frac{3}{8}}$.

13. $80 \div 2 = \mathbf{40}$

14. $\mathbf{2\frac{3}{4}}$ **in.**

15. 6×7 ⓢ $50 - 10$ because $42 > 40$

(circled: $>$)

16. Perimeter $= 10$ mm $+ 20$ mm $+ 10$ mm $+ 20$ mm
$=$ **60 mm**

17. 2, 6, 10, 14, **18**
Each number is 4 more than the preceding number.

18. **C. 4321** Note that it ends with an odd digit.

19. $\mathbf{4\overline{)20}}$ $\mathbf{20 \div 4}$ $\mathbf{\dfrac{20}{4}}$

20. **10 + 20 = 30** **20 + 10 = 30**
30 − 20 = 10 **30 − 10 = 20**

Test 2, Form B

1.
```
  $10.00
−  $7.37
  $2.63
```
$2.63

2. Two dozen is 24.
24 ÷ 8 = 3
Each person will receive **3 cookies.**

3.
```
   678
    48
+ 1386
  2112
```
2112

4.
```
  503
−  86
  417
```
417

5.
```
   34
×  36
  204
 102
 1224
```
1224

6.
```
    602
6)3612
  36
   012
    12
     0
```
602

7.
```
   $6.25
  $12.00
+  $0.59
  $18.84
```
$18.84

8. 18 − (9 + 6) = 18 − 15 = **3**

9.
```
     32
25)800
   75
    50
    50
     0
```
32

10. $46 + n = 81$
```
  81
− 46
  35
```
35

11. $6m = 306$

$m = \frac{306}{6}$
```
   51
6)306
  30
   06
    6
    0
```
51

12. Three out of ten is $\mathbf{\frac{3}{10}}$.

13. 36 ÷ 2 = **18**

14. **27 mm**

15. 12 × 7 (=) 7 × 12
Note that reversing the order of the factors does not change the product.

16. Perimeter = 10 mm × 4 = **40 mm**

17. 1, 2, 4, 8, 16, **32**
Each number is twice the preceding number.

18. **C. 7654** Note that it ends with an even digit.

19. **8)16** **16 ÷ 8** $\mathbf{\frac{16}{8}}$

20. **11 + 22 = 33** **22 + 11 = 33**
33 − 22 = 11 **33 − 11 = 22**

Test 3, Form A

1. 1000 pennies ÷ 50 pennies in each roll = **20 rolls**

2.
```
  1776
− 1215
   561 years
```
561 years

3. **ten thousands**

4. **ten million**

5. (8 + 4) ÷ (8 − 4) = 12 ÷ 4 = **3**

6.
```
   5020
×    86
  30120
 40160
431,720
```
431,720

7. $\begin{array}{r} 354 \\ -\ 189 \\ \hline \textbf{165 steps} \end{array}$

8. **12°F** (*Note:* Tick marks are spaced 2° apart.)

9. Two out of 5 is $\mathbf{\frac{2}{5}}$.

10. $-7 \ \textcircled{<} \ -4$

11. $\begin{array}{r} 5{,}874 \\ 36{,}287 \\ +\quad 489 \\ \hline \mathbf{42{,}650} \end{array}$

12. $\begin{array}{r} 48 \\ \times\ 67 \\ \hline 336 \\ 288 \\ \hline \mathbf{3216} \end{array}$

13. $\begin{array}{r} \$1.25 \\ \times \qquad 8 \\ \hline \mathbf{\$10.00} \end{array}$

14. $\begin{array}{r} \mathbf{506} \\ 7\overline{)3542} \\ 35 \\ \hline 042 \\ 42 \\ \hline 0 \end{array}$

15. $\begin{array}{r} \mathbf{25} \\ 25\overline{)625} \\ 50 \\ \hline 125 \\ 125 \\ \hline 0 \end{array}$

16. $24 \div (12 \div 4) = 24 \div 3 = \mathbf{8}$

17. $1000 - m = 408$ $\qquad \begin{array}{r} 1000 \\ -\ \ 408 \\ \hline \mathbf{592} \end{array}$

18. $9n = 63$ $\qquad \begin{array}{r} \mathbf{7} \\ 9\overline{)63} \end{array}$

$n = \frac{63}{9}$

19. Perimeter $= 4 \times 20$ mm $=$ **80 mm**

20. $\mathbf{8 \times 15 = 120} \qquad \mathbf{15 \times 8 = 120}$

$\mathbf{120 \div 8 = 15} \qquad \mathbf{120 \div 15 = 8}$

Test 3, Form B

1. 234 players ÷ 18 teams = **13 players per team**

2. $\begin{array}{r} 1776 \\ -\ 1620 \\ \hline \textbf{156 years} \end{array}$

3. **thousands**

4. **four billion**

5. $(4 \times 5) - (4 + 5) = 20 - 9 = \mathbf{11}$

6. $\begin{array}{r} \mathbf{20} \\ 12\overline{)240} \end{array}$

7. $\begin{array}{r} 354 \\ -\ 178 \\ \hline \textbf{176 steps} \end{array}$

8. **2°F** (*Note:* Tick marks are spaced 2° apart.)

9. One out of 8 is $\mathbf{\frac{1}{8}}$.

10. $8 - 5 \ \textcircled{>} \ 5 - 8$

11. $\begin{array}{r} \$3.75 \\ \$0.48 \\ +\ \$5.00 \\ \hline \mathbf{\$9.23} \end{array}$

12. $\begin{array}{r} 56 \\ \times\ 83 \\ \hline 168 \\ 448 \\ \hline \mathbf{4648} \end{array}$

13. $\begin{array}{r} \$0.35 \\ \times \quad 10 \\ \hline \mathbf{\$3.50} \end{array}$

14.
```
      402
 12)4824
    48
    024
     24
      0
```

15.
```
   $4.17
 3)$12.51
    12
     0 5
       3
      21
      21
       0
```

16. $24 \div (12 \div 6) = 24 \div 2 =$ **12**

17. $3000 - a = 1096$
```
   3000
 - 1096
   1904
```
1904

18. $9m = 54$

$m = \frac{54}{9}$
```
    6
 9)54
```
6

19. Perimeter = 2 cm + 1 cm + 2 cm + 1 cm = **6 cm**

20. **8 + 15 = 23** **15 + 8 = 23**
23 − 8 = 15 **23 − 15 = 8**

TEST 4, FORM A

1. $(9 + 3) \div (9 - 3) = 12 \div 6 =$ **2**

2. **3**

3.
```
   2 10 1
   3  1  2   (crossed out: 3, 1)
 -    9  6
   2  1  6  pages
```
216 pages

4. From −5° to 4° is **9°**.

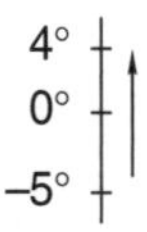

5 + 4 = 9

5. $Q - 40 = 25$
```
   25
 + 40
   65
```
65

6. $3n = 87$

$n = \frac{87}{3}$
```
    29
  3)87
    6
    27
    27
     0
```
29

7. Since the digit in the hundreds place is 5 or more, 37,542 rounds to **38,000.**

8. We estimate 51×48 by multiplying 50×50, which is **2500.**

9.
```
    41
  3)123
    12
     03
      3
      0
```
41

10. $\frac{3 + 7 + 9 + 10 + 10 + 9}{6} = \frac{48}{6} =$ **8**

11. $\frac{24 + 98}{2} = \frac{122}{2} =$ **61**

12. **1, 2, 3, 6, 9, 18** are factors (divisors) of 18.

13.
```
    $0.39
  ×    25
      195
      78
    $9.75
```
$9.75

14. Factors of 18: 1, 2, 3, 6, 9, 18
Factors of 24: 1, 2, 3, 4, 6, 8, 12, 24
The GCF of 18 and 24 is **6.**

15. **Friday**

16. **5 answers**

17. $3.50 ÷ 2 = **$1.75**

18. Perimeter = 12 mm + 18 mm + 12 mm + 18 mm = **60 mm**

19. 15, 30, 45, 60, **75**
Each number is 15 more than the preceding number.

20. $2\frac{1}{2}$ **in.**

Test 4, Form B

1. $(4 \times 3) - (3 + 4) = 12 - 7 =$ **5**

2. **5**

3. $\begin{array}{r} \overset{1}{\cancel{2}}\overset{10}{\cancel{1}}\overset{1}{1} \\ -\ 94 \\ \hline 117 \end{array}$ **pages**

4. From –8° to 10° is **18°.**

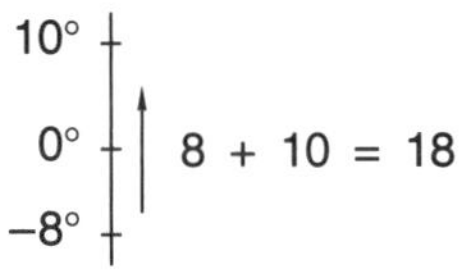

5. $R - 56 = 62$ $\qquad \begin{array}{r} 62 \\ +\ 56 \\ \hline \mathbf{118} \end{array}$

6. $3m = 78$ $\qquad \begin{array}{r} \mathbf{26} \\ 3\overline{)78} \\ 6 \\ \hline 18 \\ 18 \\ \hline 0 \end{array}$
$m = \frac{78}{3}$

7. The digit in the tens place (6) is 5 or more, so 4367 rounds to **4400.**

8. We estimate $5280 + 3910$ by adding $5000 + 4000$, which is **9000.**

9. $\begin{array}{r} \mathbf{72} \\ 6\overline{)432} \\ 42 \\ \hline 12 \\ 12 \\ \hline 0 \end{array}$

10. $\frac{6 + 10 + 8 + 7 + 9}{5} = \frac{40}{5} = \mathbf{8}$

11. $\frac{17 + 83}{2} = \frac{100}{2} = \mathbf{50}$

12. **1, 3, 7, 21** are factors (divisors) of 21.

13. $\begin{array}{r} \$0.37 \\ \times\ \ 25 \\ \hline 185 \\ 74\ \\ \hline \mathbf{\$9.25} \end{array}$

14. Factors of 16: 1, 2, 4, 8, 16
Factors of 24: 1, 2, 3, 4, 6, 8, 12, 24
The GCF of 16 and 24 is **8.**

15. $12 + 10 = 22$
Maria and Susan together scored **22 points.**

16. $9 - 7 = 2$
Tony scored **2 points** more than Robert.

17. $\$7.50 \div 2 =$ **$3.75**

18. Perimeter $= 4 \times 8$ mm $=$ **32 mm**

19. 1, 3, 9, 27, **81**
Each number is 3 times the preceding number.

20. $1\frac{1}{2}$ **in.**

Test 5, Form A

1. $(20 \times 10) - (20 + 10) = 200 - 30 =$ **170**

2. $\begin{array}{r} 28 \\ -\ 13 \\ \hline 15 \end{array}$ $\qquad \frac{\text{boys}}{\text{girls}} = \mathbf{\frac{13}{15}}$

3. **17,105,000**

4. $87 \times 31 \approx 90 \times 30 =$ **2700**

5. $2 + 4 + 3 + 7 = 16$
$1 + 6 + 3 + 2 = 12$
$4 + 3 + 7 + 3 = 17$

B. 1632

6. $\begin{array}{r} 2 \\ 5\overline{)12} \\ 10 \\ \hline 2 \end{array}$ $\qquad \frac{12}{5} = 2\mathbf{\frac{2}{5}}$

7.
$$\begin{array}{r} 37 \\ 82 \\ 63 \\ +\ 58 \\ \hline 240 \end{array} \qquad \begin{array}{r} \mathbf{60} \\ 4\overline{)240} \end{array}$$

8. **12**

9. Factors of 12: 1, 2, 3, 4, 6, 12
Factors of 18: 1, 2, 3, 6, 9, 18
The GCF of 12 and 18 is **6.**

10. $\mathbf{\frac{7}{12}}$

11.
$$\begin{array}{r} \mathbf{15} \\ 4\overline{)60} \\ \underline{4} \\ 20 \\ \underline{20} \\ 0 \end{array}$$

12. $\begin{array}{r} 4 \\ 3\overline{)12} \end{array}$ $\quad 4 \times 2 = \mathbf{8}$

13. $\frac{2}{6} + \frac{3}{6} = \mathbf{\frac{5}{6}}$

14. $\frac{11}{12} - \frac{4}{12} = \mathbf{\frac{7}{12}}$

15. $1 - \frac{3}{5} = \frac{5}{5} - \frac{3}{5} = \mathbf{\frac{2}{5}}$

16. $\frac{3}{8} + \frac{5}{8} = \frac{8}{8} = \mathbf{1}$

17. $\frac{1}{2}$ of a circle is **50%** of a circle.

18. $\frac{1}{8} + \frac{1}{8} + \frac{1}{8} = \mathbf{\frac{3}{8}}$

19. $27.50 ÷ 10 = **$2.75**

20. $\frac{1}{2}$ ⓖ $\frac{1}{4}$ (> circled)

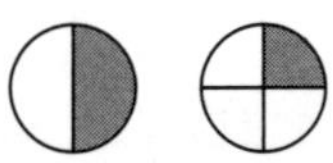

Test 5, Form B

1. (20 + 10) × (20 − 10) = 30 × 10 = **300**

2.
$$\begin{array}{r} 29 \\ -\ 13 \\ \hline 16 \end{array} \qquad \frac{\text{boys}}{\text{girls}} = \mathbf{\frac{13}{16}}$$

3. **21,286,000**

4. Since the tens digit (7) is 5 or more, 3875 rounds to **3900.**

5. 1 + 3 + 6 + 7 = 17
2 + 4 + 5 + 8 = 19
3 + 6 + 2 + 4 = 15

C. 3624

6.
$$\begin{array}{r} 5 \\ 3\overline{)16} \\ \underline{15} \\ 1 \end{array} \qquad \frac{16}{3} = \mathbf{5\frac{1}{3}}$$

7.
$$\begin{array}{r} 3{,}742 \\ 4{,}918 \\ +\ 5{,}170 \\ \hline 13{,}830 \end{array} \qquad \begin{array}{r} \mathbf{4610} \\ 3\overline{)13{,}830} \\ \underline{12} \\ 1\ 8 \\ \underline{1\ 8} \\ 03 \\ \underline{3} \\ 00 \end{array}$$

8. **7**

9. Factors of 10: 1, 2, 5, 10
Factors of 15: 1, 3, 5, 15
The GCF of 10 and 15 is **5.**

10. $\mathbf{\frac{4}{15}}$

11. $\begin{array}{r} \mathbf{20} \\ 3\overline{)60} \end{array}$

12. $\begin{array}{r} 6 \\ 4\overline{)24} \end{array}$ $\quad 6 \times 3 = \mathbf{18}$

13. $\frac{1}{3} + \frac{1}{3} = \mathbf{\frac{2}{3}}$

14. $\frac{7}{8} - \frac{4}{8} = \mathbf{\frac{3}{8}}$

15. $1 - \frac{3}{8} = \frac{8}{8} - \frac{3}{8} = \mathbf{\frac{5}{8}}$

16. $\frac{5}{6} + \frac{1}{6} = \frac{6}{6} = \mathbf{1}$

17. $\frac{1}{4}$ of a circle is **25%** of a circle.

18. $\frac{1}{12} + \frac{2}{12} + \frac{4}{12} = \mathbf{\frac{7}{12}}$

19. $37.50 ÷ 10 = **$3.75**

20. $\frac{1}{2}$ (<) $\frac{3}{4}$

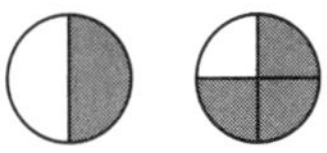

TEST 6, FORM A

1. $4\overline{)27}$ = 6, 24, remainder 3 — $\mathbf{6\frac{3}{4}}$ **inches**

$$\begin{array}{r} 6 \\ 4\overline{)27} \\ \underline{24} \\ 3 \end{array}$$

2. $$\begin{array}{r} 30 \\ \underline{-\ 16} \\ 14 \end{array}$$ $\frac{\text{boys}}{\text{girls}} = \frac{14}{16} = \mathbf{\frac{7}{8}}$

3. $68 \times 21 \approx 70 \times 20 = \mathbf{1400}$

4. $$\begin{array}{r} \mathbf{63} \\ 9\overline{)567} \\ \underline{54} \\ 27 \\ \underline{27} \\ 0 \end{array}$$

5. $$\begin{array}{r} 68 \\ 72 \\ \underline{+\ 82} \\ 222 \end{array} \qquad \begin{array}{r} \mathbf{74} \\ 3\overline{)222} \\ \underline{21} \\ 12 \\ \underline{12} \\ 0 \end{array}$$

6. The factors of 20 are **1, 2, 4, 5, 10, 20.**

7. Factors of 24: 1, 2, 3, 4, 6, 8, 12, 24
Factors of 16: 1, 2, 4, 8, 16
The GCF of 24 and 16 is **8.**

8. **B. diameter**

9. $\frac{3}{4} \times \frac{12}{1} = \frac{36}{4}$ which is **9 doughnuts.**

10. $\left(\frac{2}{7} + \frac{2}{7}\right) - \frac{4}{7} = \frac{4}{7} - \frac{4}{7} = \mathbf{0}$

11. $1 - \frac{1}{3} = \frac{3}{3} - \frac{1}{3} = \mathbf{\frac{2}{3}}$

12. $\frac{1}{2} \times \frac{3}{4} = \mathbf{\frac{3}{8}}$

13. $$\begin{array}{r} 1\frac{4}{6} \\ 6\overline{)10} \\ \underline{6} \\ 4 \end{array} \qquad 1\frac{4}{6} = \mathbf{1\frac{2}{3}}$$

14. $3\frac{2}{3} + 2\frac{2}{3} = 5\frac{4}{3} = \mathbf{6\frac{1}{3}}$

15. $\frac{3}{8} + \frac{3}{8} = \frac{6}{8} = \mathbf{\frac{3}{4}}$

16. Multiples of 4: 4, 8, 12, 16, 20, 24, …
Multiples of 6: 6, 12, 18, 24, …
The LCM of 4 and 6 is **12.**

17. 36 in. ÷ 4 = **9 in.**

18. $\frac{2}{3} \times \mathbf{\frac{3}{2}} = 1$

19. 8, 16, 24, 32, **40, 48, 56**
Each number is 8 more than the preceding number. Note that these numbers are the multiples of 8.

20. $\angle C$ or $\angle ACB$ or $\angle BCA$

Test 6, Form B

1. $\begin{array}{r} 7 \\ 4\overline{)29} \\ \underline{28} \\ 1 \end{array}$ $\mathbf{7\frac{1}{4}}$ **inches**

2. $\begin{array}{r} 30 \\ \underline{-\ 18} \\ 12 \end{array}$ $\frac{\text{boys}}{\text{girls}} = \frac{12}{18} = \mathbf{\frac{2}{3}}$

3. $6084 - 3973 \approx 6000 - 4000 = \mathbf{2000}$

4. $\begin{array}{r} \mathbf{24} \\ 18\overline{)432} \\ \underline{36} \\ 72 \\ \underline{72} \\ 0 \end{array}$

5. $\begin{array}{r} 143 \\ 98 \\ \underline{+\ 128} \\ 369 \end{array}$ $\begin{array}{r} \mathbf{123} \\ 3\overline{)369} \\ \underline{3} \\ 06 \\ \underline{6} \\ 09 \\ \underline{9} \\ 0 \end{array}$

6. The factors of 33 are **1, 3, 11, 33.**

7. Factors of 27: 1, 3, 9, 27
   Factors of 45: 1, 3, 5, 9, 15, 45
   The GCF of 27 and 45 is **9.**

8. **C. circumference**

9. $\frac{3}{4} \times \frac{20}{1} = \frac{60}{4} = \mathbf{15}$

10. $\left(\frac{3}{5} - \frac{2}{5}\right) + \frac{1}{5} = \frac{1}{5} + \frac{1}{5} = \mathbf{\frac{2}{5}}$

11. $1 - \frac{1}{4} = \frac{4}{4} - \frac{1}{4} = \mathbf{\frac{3}{4}}$

12. $\frac{1}{3} \times \frac{2}{5} = \mathbf{\frac{2}{15}}$

13. $\begin{array}{r} 1\frac{4}{8} \\ 8\overline{)12} \\ \underline{8} \\ 4 \end{array}$ $1\frac{4}{8} = \mathbf{1\frac{1}{2}}$

14. $4\frac{4}{5} + 3\frac{3}{5} = 7\frac{7}{5} = \mathbf{8\frac{2}{5}}$

15. $\frac{5}{6} - \frac{1}{6} = \frac{4}{6} = \mathbf{\frac{2}{3}}$

16. Multiples of 6: 6, 12, 18, 24, 30, …
    Multiples of 8: 8, 16, 24, 32, 40, …
    The LCM of 6 and 8 is **24.**

17. 60 cm ÷ 4 = **15 cm**

18. $\frac{3}{4} \times \mathbf{\frac{4}{3}} = 1$

19. 7, 14, 21, 28, 35, 42, **49**
    Each number is 7 more than the preceding number. Note that these numbers are the multiples of 7.

20. $\angle B$ or $\angle ABC$ or $\angle CBA$

Test 7, Form A

1. **1800**

2. $\mathbf{\frac{3}{2}}$

3. Perimeter = 4 × 8 in. = **32 in.**

4. 10 × 12 = **120 square tiles**

5. **3**

6. **0.012**

7. (5 × 100) + (6 × 1) = 500 + 6 = **506**

8. Radius $= \frac{\text{diameter}}{2} = \frac{24 \text{ in.}}{2} =$ **12 in.**

9. $0.47 = \mathbf{\frac{47}{100}}$

10. $\frac{3}{4} \text{ (<) } \frac{4}{3}$

Note that $\frac{3}{4}$ is less than 1 and $\frac{4}{3}$ is greater than 1.

11. $\frac{35}{100} = \mathbf{0.35}$

12. $\frac{3}{5} + \frac{4}{5} = \frac{7}{5} = \mathbf{1\frac{2}{5}}$

13. $\frac{2}{3} \times \frac{3}{5} = \frac{6}{15} = \mathbf{\frac{2}{5}}$

14. $18 \times 20 \div 12 = 360 \div 12 = \mathbf{30}$

15. $\frac{3}{4} \times \frac{36}{1} = \frac{108}{4} = \mathbf{27}$

16. $\frac{5}{8} - \frac{1}{8} - \frac{4}{8} - \mathbf{\frac{1}{2}}$

17. $w - 10 = 100$

$$\begin{array}{r} 100 \\ +\ 10 \\ \hline \mathbf{110} \end{array}$$

18. $3 + N = 4\frac{1}{3}$

$N = 4\frac{1}{3} - 3 = \mathbf{1\frac{1}{3}}$

19. $80\% = \frac{80}{100} = \mathbf{\frac{4}{5}}$

20. $$\begin{array}{r} 8 \\ 5\overline{)40} \\ 40 \\ \hline 0 \end{array} \qquad \begin{array}{r} 8 \\ \times\ 4 \\ \hline \mathbf{32} \end{array}$$ **answers**

Test 7, Form B

1. **5300**

2. $\mathbf{\frac{6}{5}}$

3. Perimeter $= 4 \times 10$ cm $=$ **40 cm**

4. $10 \times 15 =$ **150 square tiles**

5. **8**

6. **1.5**

7. $(5 \times 1000) + (7 \times 10) = 5000 + 70 = \mathbf{5070}$

8. Radius $= \frac{\text{diameter}}{2} = \frac{28 \text{ in.}}{2} =$ **14 in.**

9. $0.7 = \mathbf{\frac{7}{10}}$

10. $\frac{6}{5} \text{ (>) } \frac{5}{6}$

Note that $\frac{6}{5}$ is greater than 1 and $\frac{5}{6}$ is less than 1.

11. $\frac{3}{100} = \mathbf{0.03}$

12. $\frac{3}{7} + \frac{6}{7} = \frac{9}{7} = \mathbf{1\frac{2}{7}}$

13. $\frac{4}{5} \times \frac{3}{4} = \frac{12}{20} = \mathbf{\frac{3}{5}}$

14. $23 \times (500 \div 20) = 23 \times 25 = \mathbf{575}$

15. $\frac{5}{6} \times \frac{36}{1} = \frac{180}{6} = \mathbf{30}$

16. $\frac{5}{6} - \frac{1}{6} = \frac{4}{6} = \mathbf{\frac{2}{3}}$

17. $w - 50 = 500$

$$\begin{array}{r} 500 \\ +\ 50 \\ \hline \mathbf{550} \end{array}$$

18. $3 + N = 5\frac{2}{3}$

$N = 5\frac{2}{3} - 3 = \mathbf{2\frac{2}{3}}$

19. $60\% = \frac{60}{100} = \mathbf{\frac{3}{5}}$

20. $4\overline{)40}$ = 10; 40; 0

$10 \times 3 =$ **30 answers**

Test 8, Form A

1. **∠*BDC* or ∠*CDB***

2. **5**

3. $\frac{16}{20} = \mathbf{\frac{4}{5}}$

4. $6107 - 2960 \approx 6000 + 3000 = \mathbf{9000}$

5. $4^2 + \sqrt{9} = 16 + 3 = \mathbf{19}$

6. $78 + 75 + 90 = 243$

$243 \div 3 = \mathbf{81}$ (3)243: 24, 03, 3, 0

7. Factors of 12: 1, 2, 3, 4, 6, 12
Factors of 20: 1, 2, 4, 5, 10, 20
The GCF of 12 and 20 is **4.**

8. $\frac{60}{100} = \frac{6}{10} = \mathbf{\frac{3}{5}}$

9. $1\frac{2}{3} + 2\frac{2}{3} = 3\frac{4}{3} = \mathbf{4\frac{1}{3}}$

10. $1 - \frac{5}{8} = \frac{8}{8} - \frac{5}{8} = \mathbf{\frac{3}{8}}$

11. $\frac{1}{3} \times \frac{3}{4} = \frac{3}{12} = \mathbf{\frac{1}{4}}$

12. **1 hr 53 min**

13. Multiples of 6: 6, 12, 18, 24, 30, …
Multiples of 8: 8, 16, 24, 32, 40, …
LCM of 6 and 8 is **24.**

14. $10 \times 10 =$ **100 square-inch tiles**

15. **3**

16. **10.05**

17. $0.7 + 0.8 = \mathbf{1.5}$

18. $0.2 + 0.34 + 5.0 = \mathbf{5.54}$

19. $0.15 - 0.06 = \mathbf{0.09}$

20. $0.15 \times 0.3 = \mathbf{0.045}$

Test 8, Form B

1. **∠*ADB* or ∠*BDA***

2. **2**

3. $\frac{15}{20} = \mathbf{\frac{3}{4}}$

4. $8972 + 4106 \approx 9000 + 4000 = \mathbf{13{,}000}$

5. $5^2 + \sqrt{16} = 25 + 4 = \mathbf{29}$

6. $81 + 84 + 84 = 249$

$249 \div 3 = \mathbf{83}$ (3)249: 24, 09, 9, 0

7. Factors of 21: 1, 3, 7, 21
Factors of 35: 1, 5, 7, 35
The GCF of 21 and 35 is **7.**

8. $\frac{40}{100} = \frac{4}{10} = \mathbf{\frac{2}{5}}$

9. $3\frac{3}{4} + 1\frac{1}{4} = 4\frac{4}{4} = \mathbf{5}$

10. $5 - 1\frac{1}{3} = 4\frac{3}{3} - 1\frac{1}{3} = \mathbf{3\frac{2}{3}}$

11. $\frac{1}{2} \times \frac{2}{3} = \frac{2}{6} = \mathbf{\frac{1}{3}}$

12. **6 hr 15 min**

13. Multiples of 3: 3, 6, 9, …
Multiples of 6: 6, 12, 18, …
LCM of 3 and 6 is **6.**

14. 6 × 8 = **48 square-inch tiles**

15. **4**

16. **3.12**

17. $$\begin{array}{r} 0.3 \\ 0.4 \\ +\ 0.5 \\ \hline \mathbf{1.2} \end{array}$$

18. $$\begin{array}{r} 6.0 \\ 4.2 \\ +\ 0.36 \\ \hline \mathbf{10.56} \end{array}$$

19. $$\begin{array}{r} 0.23 \\ -\ 0.19 \\ \hline \mathbf{0.04} \end{array}$$

20. $$\begin{array}{r} 0.12 \\ \times\ \ 0.4 \\ \hline \mathbf{0.048} \end{array}$$

Test 9, Form A

1. (10 × 6) − (7 + 8) = 60 − 15 = **45**

2. $\frac{3}{4} \times \frac{4}{3} = 1;\ \frac{4}{3} = \mathbf{1\frac{1}{3}}$

3. 36 in. ÷ 4 = **9 in.**

4. Area = 12 ft × 15 ft = **180 sq. ft**

5. **10.12**

6. $\frac{25}{100} = \frac{1}{4} \qquad \frac{1}{4} \times \frac{\$80}{1} = \mathbf{\$20}$

7. $\frac{3}{4} \times \frac{24}{1} = \frac{72}{4} = \mathbf{18}$

8. **C. 70%**

9. **4**

10. $7\% = \mathbf{\frac{7}{100}} = \mathbf{0.07}$

11. Factors of 12: 1, 2, 3, 4, 6, 12
Factors of 8: 1, 2, 4, 8
The GCF of 12 and 8 is **4.**

12. $2\frac{1}{3} + \sqrt{9} = 2\frac{1}{3} + 3 = \mathbf{5\frac{1}{3}}$

13. $\frac{1}{2} \times \frac{2}{3} = \frac{2}{6} = \mathbf{\frac{1}{3}}$

14. $$\begin{array}{r} 0.43 \\ 6.7 \\ +\ 8.0 \\ \hline \mathbf{15.13} \end{array}$$

15. $$\begin{array}{r} 5.00 \\ -\ 3.27 \\ \hline \mathbf{1.73} \end{array}$$

16. $$\begin{array}{r} 0.34 \\ \times\ 0.26 \\ \hline 204 \\ 68 \\ \hline \mathbf{0.0884} \end{array}$$

17. $4\frac{2}{3} - 3\frac{1}{3} = \mathbf{1\frac{1}{3}}$

18. $\frac{2}{3} \times \frac{4}{4} = \mathbf{\frac{8}{12}}$

19.
$$\begin{array}{r} 0.175 \\ 4\overline{)0.700} \\ \underline{4} \\ 30 \\ \underline{28} \\ 20 \\ \underline{20} \\ 0 \end{array}$$

20. C. $\overline{SR}$

TEST 9, FORM B

1. $(8 + 2) \times (8 - 2) = 10 \times 6 = \mathbf{60}$

2. $\frac{2}{3} \times \frac{3}{2} = 1; \frac{3}{2} = \mathbf{1\frac{1}{2}}$

3. $100 \text{ cm} \div 4 = \mathbf{25\ cm}$

4. Area = 11 ft × 12 ft = **132 sq. ft**

5. **12.05**

6. $\frac{25}{100} = \frac{1}{4}$ $\quad \frac{1}{4} \times \frac{\$60}{1} = \mathbf{\$15}$

7. $\frac{2}{3} \times \frac{24}{1} = \frac{48}{3} = \mathbf{16}$

8. **A. 30%**

9. **7**

10. $9\% = \mathbf{\frac{9}{100}} = \mathbf{0.09}$

11. Multiples of 12: 12, 24, 36, 48, …
Multiples of 8: 8, 16, 24, 32, 40, 48, …
The LCM of 12 and 8 is **24.**

12. $1\frac{3}{5} + \sqrt{4} = 1\frac{3}{5} + 2 = \mathbf{3\frac{3}{5}}$

13. $\frac{1}{3} \times \frac{3}{4} = \frac{3}{12} = \mathbf{\frac{1}{4}}$

14.
$$\begin{array}{r} 6.2 \\ 7.0 \\ +\ 0.48 \\ \hline \mathbf{13.68} \end{array}$$

15.
$$\begin{array}{r} 2.00 \\ -\ 0.34 \\ \hline \mathbf{1.66} \end{array}$$

16.
$$\begin{array}{r} 0.23 \\ \times\ 0.12 \\ \hline 46 \\ 23 \\ \hline \mathbf{0.0276} \end{array}$$

17. $3\frac{3}{5} + 3\frac{2}{5} = 6\frac{5}{5} = \mathbf{7}$

18. $\frac{2}{3} \times \frac{3}{3} = \mathbf{\frac{6}{9}}$

19.
$$\begin{array}{r} 0.375 \\ 8\overline{)3.000} \\ \underline{2\,4} \\ 60 \\ \underline{56} \\ 40 \\ \underline{40} \\ 0 \end{array}$$

20. D. $\overline{SQ}$

TEST 10, FORM A

1. $(0.3 + 0.4) - (0.3 \times 0.4) = 0.7 - 0.12 = \mathbf{0.58}$

2. **3**

3. $8\% = \frac{8}{100} = 0.08$
$$\begin{array}{r} \$16 \\ \times\ 0.08 \\ \hline \mathbf{\$1.28} \end{array}$$

4. $5867 + 2106 \approx 6000 + 2000 = \mathbf{8000}$

5.
$$\begin{array}{r} \mathbf{2005} \\ 20\overline{)40{,}100} \end{array}$$

6.
$$\begin{array}{r} 96 \\ 100 \\ +\ 92 \\ \hline 288 \end{array} \qquad \begin{array}{r} \mathbf{96} \\ 3\overline{)288} \\ \underline{27} \\ 18 \\ \underline{18} \\ 0 \end{array}$$

7. $\frac{7}{24} + \frac{11}{24} = \frac{18}{24} = \mathbf{\frac{3}{4}}$

8. $\frac{5}{6} \times \frac{24}{1} = \frac{120}{6} = \mathbf{20}$

9. $\frac{1}{4} = \frac{2}{8} \qquad \frac{2}{8} + \frac{3}{8} = \mathbf{\frac{5}{8}}$

10. Multiples of 2: 2, 4, 6, 8, 10, 12, …
Multiples of 3: 3, 6, 9, 12, 15, 18, …
Multiples of 4: 4, 8, 12, 16, 20, 24, …
The LCM of 2, 3, and 4 is **12.**

11. Area = 100 cm × 100 cm = **10,000 sq. cm**

12.
$$\begin{array}{r} 3.14 \\ \times\ \ 10 \\ \hline 0 \\ 3140 \\ \hline \end{array}$$
31.40 in. or **31.4 in.**

13. **10.2**

14.
$$\begin{array}{r} 1.00 \\ -\ 0.23 \\ \hline \mathbf{0.77} \end{array}$$

15.
$$\begin{array}{r} 5.62 \\ 4.0 \\ +\ 0.5 \\ \hline \mathbf{10.12} \end{array}$$

16. 0.2 × 0.3 × 0.4 = **0.024**

17.
$$\begin{array}{r} \mathbf{0.041} \\ 3\overline{)0.123} \end{array}$$

18.
$$\begin{array}{r} \mathbf{1.5} \\ 0.3\overline{)0.45} \\ \underline{3} \\ 15 \\ \underline{15} \\ 0 \end{array}$$

19.
$$\begin{array}{r} 6\frac{1}{5} = 5\frac{6}{5} \\ -\ 1\frac{3}{5} = 1\frac{3}{5} \\ \hline \mathbf{4\frac{3}{5}} \end{array}$$

20. 12.5 × 100 = **1250**

Test 10, Form B

1. $\left(\frac{1}{3} + \frac{2}{3}\right) - \left(\frac{1}{3} \times \frac{2}{3}\right) = 1 - \frac{2}{9} = \mathbf{\frac{7}{9}}$

2. **7**

3. $8\% = \frac{8}{100} = 0.08 \qquad \begin{array}{r} \$14 \\ \times\ 0.08 \\ \hline \mathbf{\$1.12} \end{array}$

4. 48 × 21 ≈ 50 × 20 = **1000**

5.
$$\begin{array}{r} \mathbf{209} \\ 12\overline{)2508} \\ \underline{24}\ \ \\ 108 \\ \underline{108} \\ 0 \end{array}$$

6.
$$\begin{array}{r} 47 \\ 38 \\ 56 \\ +\ 43 \\ \hline 184 \end{array} \qquad \begin{array}{r} \mathbf{46} \\ 4\overline{)184} \\ \underline{16}\ \\ 24 \\ \underline{24} \\ 0 \end{array}$$

7. $\frac{11}{36} + \frac{13}{36} = \frac{24}{36} = \mathbf{\frac{2}{3}}$

8. $\frac{3}{4} \times \frac{48}{1} = \frac{144}{4} = \mathbf{36}$

9. $\frac{1}{4} = \frac{2}{8} \qquad \frac{2}{8} + \frac{5}{8} = \mathbf{\frac{7}{8}}$

10. Multiples of 2: 2, 4, 6, 8, 10, 12, …
Multiples of 4: 4, 8, 12, 16, 20, 24, …
Multiples of 8: 8, 16, 24, 32, 40, 48, …
The LCM of 2, 4, and 8 is **8.**

11. Area = 27 mm × 43 mm = **1161 sq. mm**

12.
$$\begin{array}{r} 3.14 \\ \times\ \ 10 \\ \hline 0 \\ 3140 \\ \hline \end{array}$$
31.40 ft or **31.4 ft**

13. **21.05**

14.
$$\begin{array}{r} 1.00 \\ -\ 0.01 \\ \hline \mathbf{0.99} \end{array}$$

15.
$$\begin{array}{r} 4.6 \\ 0.47 \\ +\ 12.0 \\ \hline \mathbf{17.07} \end{array}$$

16.
$$\begin{array}{r} 0.12 \\ \times\ 0.13 \\ \hline 36 \\ 12 \\ \hline \mathbf{0.0156} \end{array}$$

17.
$$\begin{array}{r} \mathbf{0.076} \\ 6\overline{)0.456} \\ 42 \\ \hline 36 \\ 36 \\ \hline 0 \end{array}$$

18.
$$\begin{array}{r} \mathbf{4.1} \\ 0.03\overline{)0.123} \\ 12 \\ \hline 03 \\ 3 \\ \hline 0 \end{array}$$

19.
$$\begin{array}{rcl} 5\frac{1}{3} & = & 4\frac{4}{3} \\ -\ 2\frac{2}{3} & = & 2\frac{2}{3} \\ \hline & & \mathbf{2\frac{2}{3}} \end{array}$$

20. 3.5 × 100 = **350**

TEST 11, FORM A

1.
$$\begin{array}{r} 3.14 \\ \times\ \ \ 8 \\ \hline \mathbf{25.12\ cm} \end{array}$$

2.
$$\begin{array}{r} 1.5 \\ \times\ 0.025 \\ \hline 75 \\ 30 \\ \hline \mathbf{0.0375} \end{array}$$

3. Perimeter = 20 mm + 24 mm + 20 mm + 24 mm = **88 mm**

4. Area = 24 mm × 20 mm = **480 sq. mm**

5. $\frac{\text{length}}{\text{width}} = \frac{24}{20} = \mathbf{\frac{6}{5}}$

6. $30\% = \frac{30}{100} = \frac{3}{10}$

$\frac{3}{10} \times \frac{50}{1} = \frac{150}{10} =$ **15 questions**

7. $7 - n = 4.7$
$$\begin{array}{r} 7.0 \\ -\ 4.7 \\ \hline \mathbf{2.3} \end{array}$$

8. 0.60 ⊜ 0.6

9. **80.08**

10. 2.8735 ≈ **2.9**

11.
$$\begin{array}{rcl} 3\frac{2}{3} & & \\ +\ 1\frac{2}{3} & & \\ \hline 4\frac{4}{3} & = & \mathbf{5\frac{1}{3}} \end{array}$$

12.
$$\begin{array}{rcl} 6\frac{1}{3} & = & 5\frac{4}{3} \\ -\ 4\frac{2}{3} & = & 4\frac{2}{3} \\ \hline & & \mathbf{1\frac{2}{3}} \end{array}$$

13. $\frac{3}{8} \times \frac{4}{9} = \frac{12}{72} = \mathbf{\frac{1}{6}}$

14. $x = 1\frac{1}{3} - \frac{2}{3} = \frac{4}{3} - \frac{2}{3} = \mathbf{\frac{2}{3}}$

15. $3.75 \div 10^2 = 3.75 \div 100 =$ **0.0375**

16.
```
        0.0375
0.4)0.01500
        12
         30
         28
          20
          20
           0
```

17. $\sqrt{36} \div 0.6 = 6 \div 0.6$ $\quad 0.6\overline{)6.0}$ = **10**

18.
$$\begin{array}{rcl} \frac{3}{4} & = & \frac{6}{8} \\ + \frac{1}{8} & = & \frac{1}{8} \\ \hline & & \mathbf{\frac{7}{8}} \end{array}$$

19. $\frac{\cancel{2} \cdot 2 \cdot \cancel{3} \cdot 3 \cdot \cancel{5}}{\cancel{2} \cdot \cancel{3} \cdot \cancel{5} \cdot 7} = \frac{2 \cdot 3}{7} = \mathbf{\frac{6}{7}}$

20. **3.7**

Test 11, Form B

1.
$$\begin{array}{r} 3.14 \\ \times \quad 6 \\ \hline \mathbf{18.84\ cm} \end{array}$$

2.
$$\begin{array}{r} 0.81 \\ \times\ 0.012 \\ \hline 162 \\ 81\ \\ \hline \mathbf{0.00972} \end{array}$$

3. Perimeter = 20 mm + 15 mm + 20 mm + 15 mm = **70 mm**

4. Area = 15 mm × 20 mm = **300 sq. mm**

5. $\frac{\text{length}}{\text{width}} = \frac{20}{15} = \mathbf{\frac{4}{3}}$

6. $20\% = \frac{20}{100} = \frac{1}{5}$

$\frac{1}{5} \times \frac{50}{1} = \frac{50}{5} =$ **10 questions**

7. $6 - w = 3.1$
$$\begin{array}{r} 6.0 \\ -\ 3.1 \\ \hline \mathbf{2.9} \end{array}$$

8. $1.23 \; \textcircled{>} \; 0.140$

9. **20.2**

10. $6.7347 \approx$ **6.73**

11.
$$\begin{array}{r} 3\frac{3}{4} \\ +\ 2\frac{1}{4} \\ \hline 5\frac{4}{4} = \mathbf{6} \end{array}$$

12.
$$\begin{array}{rcl} 5\frac{1}{5} & = & 4\frac{6}{5} \\ -\ 1\frac{3}{5} & = & 1\frac{3}{5} \\ \hline & & \mathbf{3\frac{3}{5}} \end{array}$$

13. $\frac{3}{4} \cdot \frac{2}{5} = \frac{6}{20} = \mathbf{\frac{3}{10}}$

14. $x = 1\frac{2}{5} - \frac{3}{5} = \frac{7}{5} - \frac{3}{5} = \mathbf{\frac{4}{5}}$

15. $0.45 \div 10^2 = 0.45 \div 100 =$ **0.0045**

16.
```
       0.075
0.2)0.0150
       14
        10
        10
         0
```

17. $\sqrt{4} \div 0.4 = 2 \div 0.4$ $\quad 0.4\overline{)2.0}$ = **5**

18.
$$\begin{array}{rcl} \frac{1}{4} & = & \frac{2}{8} \\ + \frac{3}{8} & = & \frac{3}{8} \\ \hline & & \mathbf{\frac{5}{8}} \end{array}$$

19. $$\frac{\cancel{2} \cdot \cancel{3} \cdot 3 \cdot \cancel{5}}{\cancel{2} \cdot 2 \cdot \cancel{3} \cdot \cancel{5} \cdot 5} = \frac{3}{2 \cdot 5} = \mathbf{\frac{3}{10}}$$

20. **0.8**

Test 12, Form A

1. Out of the six possible outcomes, only 5 and 6 are greater than 4.

$$\frac{2}{6} = \mathbf{\frac{1}{3}}$$

2. Perimeter = 6 × 8 in. = **48 in.**

3. 24 cm × 4 cm = **96 sq. cm**

4. **3**

5.
$$\begin{array}{r} 3.0 \\ 4.5 \\ +\ 0.23 \\ \hline \mathbf{7.73} \end{array}$$

6.
$$\begin{array}{r} 1.00 \\ -\ 0.29 \\ \hline \mathbf{0.71} \end{array}$$

7.
$$\begin{array}{r} 0.27 \\ \times\ 0.3 \\ \hline \mathbf{0.081} \end{array}$$

8.
$$\begin{array}{r} 3.14 \\ \times\ 30 \\ \hline 94.20 \end{array} \approx \mathbf{94\ inches}$$

9. **D. 9**

10. $$\frac{\text{win}}{\text{loss}} = \frac{10}{20} = \mathbf{\frac{1}{2}}$$

11.
$$\begin{array}{rcl} 5 & = & 4\frac{3}{3} \\ -\ 2\frac{1}{3} & = & 2\frac{1}{3} \\ \hline & & \mathbf{2\frac{2}{3}} \end{array}$$

12. $$\begin{array}{r} \mathbf{0.05} \\ 9\overline{)0.45} \end{array}$$

13. $$\begin{array}{r} \mathbf{30} \\ 0.04\overline{)1.20} \end{array}$$

14. 12.34 × 10 = **123.4**

15.
$$\begin{array}{rcl} \frac{1}{2} & = & \frac{5}{10} \\ + \frac{3}{5} & = & \frac{6}{10} \\ \hline & & \frac{11}{10} = \mathbf{1\frac{1}{10}} \end{array}$$

16.
$$\begin{array}{rcl} \frac{4}{5} & = & \frac{16}{20} \\ - \frac{1}{4} & = & \frac{5}{20} \\ \hline & & \mathbf{\frac{11}{20}} \end{array}$$

17. $\frac{3}{4} \textcircled{>} \frac{2}{3}$ because $\frac{9}{12} > \frac{8}{12}$

18.
$$\begin{array}{rcl} 3\frac{1}{2} & = & 3\frac{3}{6} \\ +\ 2\frac{2}{3} & = & 2\frac{4}{6} \\ \hline & & 5\frac{7}{6} = \mathbf{6\frac{1}{6}} \end{array}$$

19.
$$\begin{array}{rcl} 7\frac{1}{4} & = & 6\frac{5}{4} \\ -\ 1\frac{3}{4} & = & 1\frac{3}{4} \\ \hline & & 5\frac{2}{4} = \mathbf{5\frac{1}{2}} \end{array}$$

20.
$$\begin{array}{r} \$8.29 \\ \times\ 0.08 \\ \hline 0.6632 \end{array}$$

The tax was **$0.66.**

TEST 12, FORM B

1. Out of the six possible outcomes, only 3 are less than 4.

$\frac{3}{6} = \mathbf{\frac{1}{2}}$

2. Perimeter = 6 × 7 in. = **42 in.**

3. 9 cm × 12 cm = **108 sq. cm**

4. **8**

5.
$$\begin{array}{r} 6.2 \\ 0.56 \\ +\ 12.0 \\ \hline \mathbf{18.76} \end{array}$$

6.
$$\begin{array}{r} 2.50 \\ -\ 0.25 \\ \hline \mathbf{2.25} \end{array}$$

7.
$$\begin{array}{r} 1.4 \\ \times\ 0.06 \\ \hline \mathbf{0.084} \end{array}$$

8.
$$\begin{array}{r} 3.14 \\ \times\ \ 50 \\ \hline 157.00 \end{array} \approx \mathbf{157\ inches}$$

9. **C. 4**

10. $\frac{\text{win}}{\text{loss}} = \frac{12}{18} = \mathbf{\frac{2}{3}}$

11.
$$\begin{array}{rcl} 1 & = & \frac{8}{8} \\ -\ \frac{5}{8} & = & \frac{5}{8} \\ \hline & & \mathbf{\frac{3}{8}} \end{array}$$

12.
$$\begin{array}{r} \mathbf{0.24} \\ 5\overline{)1.20} \\ \underline{1\ 0} \\ 20 \\ \underline{20} \\ 0 \end{array}$$

13.
$$\begin{array}{r} \mathbf{4\ 0} \\ 0.6\overline{)24.0} \end{array}$$

14. 12.34 ÷ 10 = **1.234**

15.
$$\begin{array}{rcl} \frac{3}{4} & = & \frac{9}{12} \\ +\ \frac{1}{3} & = & \frac{4}{12} \\ \hline & & \frac{13}{12} = \mathbf{1\frac{1}{12}} \end{array}$$

16.
$$\begin{array}{rcl} \frac{2}{3} & = & \frac{10}{15} \\ -\ \frac{1}{5} & = & \frac{3}{15} \\ \hline & & \mathbf{\frac{7}{15}} \end{array}$$

17. $\frac{5}{8}$ ⓧ$\frac{3}{4}$ because $\frac{5}{8} < \frac{6}{8}$

18.
$$\begin{array}{rcl} 4\frac{3}{4} & = & 4\frac{3}{4} \\ +\ 2\frac{1}{2} & = & 2\frac{2}{4} \\ \hline & & 6\frac{5}{4} = \mathbf{7\frac{1}{4}} \end{array}$$

19.
$$\begin{array}{rcl} 8\frac{1}{3} & = & 7\frac{4}{3} \\ -\ 2\frac{2}{3} & = & 2\frac{2}{3} \\ \hline & & \mathbf{5\frac{2}{3}} \end{array}$$

20.
$$\begin{array}{r} \$9.49 \\ \times\ \ 0.08 \\ \hline 0.7592 \end{array}$$

The tax was **$0.76.**

TEST 13, FORM A

1. Side length = 8 ft ÷ 4 = 2 ft
Area = 2 ft × 2 ft = **4 sq. ft**

2. $\frac{2}{3} \times \mathbf{\frac{4}{4}} = \frac{8}{12}$

3. $\frac{2}{6} = \mathbf{\frac{1}{3}}$

4. $\frac{\text{radius}}{\text{diameter}} = \frac{21}{42} = \mathbf{\frac{1}{2}}$

5. **D.** 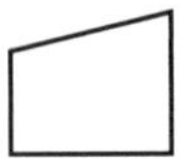

6. $5\overline{)5}$ quotient 1; $2\overline{)10}$; $2\overline{)20}$; $2\overline{)40}$

$40 = \mathbf{2 \cdot 2 \cdot 2 \cdot 5}$

7. $\frac{3}{5} \textcircled{<} \frac{2}{3}$ because $\frac{9}{15} < \frac{10}{15}$

8. $\begin{array}{r} \$12.95 \\ \times \quad 0.08 \\ \hline \$1.0360 \end{array}$ $\qquad \begin{array}{r} \$12.95 \\ + \quad \$1.04 \\ \hline \mathbf{\$13.99} \end{array}$

9. $\begin{array}{r} \$6.00 \\ - \ \$0.06 \\ \hline \mathbf{\$5.94} \end{array}$

10. $\begin{array}{r} 3.7 \\ 12.0 \\ + \ 4.31 \\ \hline \mathbf{20.01} \end{array}$

11. $\begin{array}{r} 7.40 \\ - \ 4.74 \\ \hline \mathbf{2.66} \end{array}$

12. $\begin{array}{r} 0.18 \\ \times \ 0.27 \\ \hline 126 \\ 36 \\ \hline \mathbf{0.0486} \end{array}$

13. $\begin{array}{r} \mathbf{0.028} \\ 5\overline{)0.140} \\ 10 \\ \hline 40 \\ 40 \\ \hline 0 \end{array}$

14. $\begin{array}{r} \mathbf{15} \\ 0.6\overline{)9.0} \end{array}$

15. $\begin{array}{r} 3\frac{5}{6} = 3\frac{5}{6} \\ - \ 1\frac{1}{2} = 1\frac{3}{6} \\ \hline 2\frac{2}{6} = \mathbf{2\frac{1}{3}} \end{array}$

16. $\begin{array}{r} 1\frac{2}{3} = 1\frac{4}{6} \\ + \ 1\frac{5}{6} = 1\frac{5}{6} \\ \hline 2\frac{9}{6} = 3\frac{3}{6} = \mathbf{3\frac{1}{2}} \end{array}$

17. $\begin{array}{r} \frac{1}{2} = \frac{4}{8} \\ \frac{3}{4} = \frac{6}{8} \\ + \ \frac{3}{8} = \frac{3}{8} \\ \hline \frac{13}{8} = \mathbf{1\frac{5}{8}} \end{array}$

18. $m = 6\frac{1}{3} - 3 = \mathbf{3\frac{1}{3}}$

19. **8 vertices**

20. $3\frac{3}{4} = \frac{15}{4} \qquad \frac{15}{4} \cdot \frac{1}{6} = \frac{15}{24} = \mathbf{\frac{5}{8}}$

Test 13, Form B

1. Side length = 16 in. ÷ 4 = 4 in.
Area = 4 in. × 4 in. = **16 sq. in.**

2. $\frac{3}{4} \times \mathbf{\frac{3}{3}} = \frac{9}{12}$

3. $\frac{4}{6} = \mathbf{\frac{2}{3}}$

4. $\frac{\text{radius}}{\text{diameter}} = \frac{18}{36} = \mathbf{\frac{1}{2}}$

5. **D.** 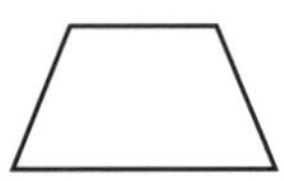

6. 50 → 5, 10; 10 → 2, 5 or 50 → 2, 25; 25 → 5, 5 $50 = \mathbf{2 \cdot 5 \cdot 5}$

7. $\frac{5}{6}$ ⓐ $\frac{5}{8}$ because $\frac{20}{24} > \frac{15}{24}$

(circled symbol: >)

8. $$\begin{array}{r} \$14.85 \\ \times \quad 0.07 \\ \hline \$1.0395 \end{array} \qquad \begin{array}{r} \$14.85 \\ + \quad \$1.04 \\ \hline \mathbf{\$15.89} \end{array}$$

9. $$\begin{array}{r} \$10.00 \\ - \quad \$0.10 \\ \hline \mathbf{\$9.90} \end{array}$$

10. $$\begin{array}{r} 4.87 \\ 16.0 \\ +\ 12.8 \\ \hline \mathbf{33.67} \end{array}$$

11. $$\begin{array}{r} 4.70 \\ -\ 1.79 \\ \hline \mathbf{2.91} \end{array}$$

12. $0.5 \times 0.6 \times 0.7 = 0.210 = \mathbf{0.21}$

13. $$\begin{array}{r} \mathbf{0.0375} \\ 4\overline{)0.1500} \\ 12 \\ \hline 30 \\ 28 \\ \hline 20 \\ 20 \\ \hline 0 \end{array}$$

14. $$\begin{array}{r} \mathbf{25} \\ 0.6\overline{)15.0} \\ 12 \\ \hline 30 \\ 30 \\ \hline 0 \end{array}$$

15. $$\begin{array}{r} 4\frac{2}{3} = 4\frac{4}{6} \\ -\ 1\frac{1}{6} = 1\frac{1}{6} \\ \hline 3\frac{3}{6} = \mathbf{3\frac{1}{2}} \end{array}$$

16. $$\begin{array}{r} 3\frac{1}{2} = 3\frac{3}{6} \\ +\ 4\frac{5}{6} = 4\frac{5}{6} \\ \hline 7\frac{8}{6} = \mathbf{8\frac{1}{3}} \end{array}$$

17. $$\begin{array}{r} \frac{1}{2} = \frac{3}{6} \\ \frac{1}{3} = \frac{2}{6} \\ +\ \frac{1}{6} = \frac{1}{6} \\ \hline \frac{6}{6} = \mathbf{1} \end{array}$$

18. $n = 5\frac{1}{4} - 2 = \mathbf{3\frac{1}{4}}$

19. **12 edges**

20. $2\frac{2}{3} = \frac{8}{3} \qquad \frac{8}{3} \cdot \frac{1}{4} = \frac{8}{12} = \mathbf{\frac{2}{3}}$

Test 14, Form A

1. **9**

2. Side length = 44 cm ÷ 4 = 11 cm
Area = 11 cm × 11 cm = **121 sq. cm**

3. $10 - (0.25 + 1) = 10 - 1.25 = \mathbf{8.75}$

4. $$\begin{array}{r} 0.12 \\ \times\ 0.12 \\ \hline 24 \\ 120 \\ \hline \mathbf{0.0144} \end{array}$$

5. $\frac{11}{20} \cdot \frac{5}{5} = \mathbf{\frac{55}{100}}$

6. $0.45 \div \sqrt{36} = 0.45 \div 6$ $$\begin{array}{r} \mathbf{0.075} \\ 6\overline{)0.450} \\ 42 \\ \hline 30 \\ 30 \\ \hline 0 \end{array}$$

7. **C.** 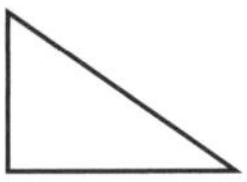

8. 0.2 + 0.2 (>) 0.2 × 0.2 because 0.4 > 0.04

9.

$$\begin{array}{r} \$24.95 \\ \times\ \ 0.06 \\ \hline \$1.4970 \end{array} \qquad \begin{array}{r} \$24.95 \\ +\ \ \$1.50 \\ \hline \mathbf{\$26.45} \end{array}$$

10. $4.5 \times 100 = \mathbf{450}$

11.

$$\begin{array}{r} \mathbf{15} \\ 0.08\overline{)1.20} \\ \underline{8} \\ 40 \\ \underline{40} \\ 0 \end{array}$$

12.

$$\begin{array}{rcl} 4\frac{1}{5} & = & 3\frac{12}{10} \\ -\ 3\frac{1}{2} & = & 3\frac{5}{10} \\ \hline & & \mathbf{\frac{7}{10}} \end{array}$$

13.

$$\begin{array}{rcl} 1\frac{1}{2} & = & 1\frac{4}{8} \\ 2\frac{1}{4} & = & 2\frac{2}{8} \\ +\ 3\frac{3}{8} & = & 3\frac{3}{8} \\ \hline & & 6\frac{9}{8} = \mathbf{7\frac{1}{8}} \end{array}$$

14. $2 \times 2 \times 2 = $ **8 small cubes**

15. Perimeter = 12 mm × 6 = **72 mm**

16. $\frac{24}{30} = \frac{\mathbf{2 \cdot 2 \cdot 2 \cdot 3}}{\mathbf{2 \cdot 3 \cdot 5}} = \mathbf{\frac{4}{5}}$

17. $2\frac{1}{4} \times 1\frac{2}{3} = \frac{\overset{3}{\cancel{9}}}{4} \times \frac{5}{\underset{1}{\cancel{3}}} = \frac{15}{4} = \mathbf{3\frac{3}{4}}$

18. $1\frac{3}{4} \div 2 = \frac{7}{4} \times \frac{1}{2} = \mathbf{\frac{7}{8}}$

19. A line segment has two endpoints.

C.

20. $\frac{\text{red}}{\text{total}} = \frac{3}{12} = \mathbf{\frac{1}{4}}$

Test 14, Form B

1. **4**

2. Side length = 48 mm ÷ 4 = 12 mm
Area = 12 mm × 12 mm = **144 sq. mm**

3. $2 - (0.6 + 0.7) = 2 - 1.3 = \mathbf{0.7}$

4.

$$\begin{array}{r} 0.12 \\ \times\ 0.13 \\ \hline 36 \\ 12\ \\ \hline \mathbf{0.0156} \end{array}$$

5. $\frac{9}{20} \cdot \frac{5}{5} = \mathbf{\frac{45}{100}}$

6. $0.28 \div \sqrt{64} = 0.28 \div 8$

$$\begin{array}{r} \mathbf{0.035} \\ 8\overline{)0.280} \\ \underline{24} \\ 40 \\ \underline{40} \\ 0 \end{array}$$

7. **C.** 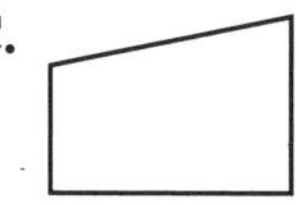

8. 0.2 + 0.2 (<) 0.2 ÷ 0.2 because 0.4 < 1

9.

$$\begin{array}{r} \$18.97 \\ \times\ \ 0.08 \\ \hline \$1.5176 \end{array} \qquad \begin{array}{r} \$18.97 \\ +\ \ \$1.52 \\ \hline \mathbf{\$20.49} \end{array}$$

10. $4.5 \div 100 = \mathbf{0.045}$

11.

$$\begin{array}{r} \mathbf{7.5} \\ 0.4\overline{)3.00} \\ \underline{28} \\ 20 \\ \underline{20} \\ 0 \end{array}$$

12.
$$\begin{array}{r} 5\frac{1}{8} = 4\frac{9}{8} \\ -\ 3\frac{1}{2} = 3\frac{4}{8} \\ \hline \mathbf{1\frac{5}{8}} \end{array}$$

13.
$$\begin{array}{r} 6\frac{1}{3} = 6\frac{2}{6} \\ 2\frac{1}{2} = 2\frac{3}{6} \\ +\ 5\frac{1}{6} = 5\frac{1}{6} \\ \hline 13\frac{6}{6} = \mathbf{14} \end{array}$$

14. $3 \times 3 \times 3 =$ **27 small cubes**

15. 48 cm $\div$ 8 = **6 cm**

16. $\frac{24}{60} = \frac{\mathbf{2 \cdot 2 \cdot 2 \cdot 3}}{\mathbf{2 \cdot 2 \cdot 3 \cdot 5}} = \mathbf{\frac{2}{5}}$

17. $1\frac{1}{3} \times 1\frac{3}{4} = \frac{\overset{1}{\cancel{4}}}{3} \times \frac{7}{\underset{1}{\cancel{4}}} = \frac{7}{3} = \mathbf{2\frac{1}{3}}$

18. $1\frac{1}{2} \div 2\frac{1}{2} = \frac{3}{2} \times \frac{2}{5} = \mathbf{\frac{3}{5}}$

19. A ray has one endpoint.

B.

20. $\frac{\text{white}}{\text{total}} = \frac{4}{12} = \mathbf{\frac{1}{3}}$

TEST 15, FORM A

1. 6 cm $\cdot$ 4 cm = **24 sq. cm**

2.
$$\begin{array}{r} \$6.80 \\ \times\ \ 0.07 \\ \hline \$0.4760 \end{array} \qquad \begin{array}{r} \$6.80 \\ +\ \$0.48 \\ \hline \$7.28 \end{array} \qquad \begin{array}{r} \$10.00 \\ -\ \ \$7.28 \\ \hline \mathbf{\$2.72} \end{array}$$

3. $\frac{3}{4}$ ⓖ $\frac{11}{16}$ because $\frac{12}{16} > \frac{11}{16}$

4. **Point *D***

5. **(3, –2)**

6. **B. pentagon**

7. $\frac{45}{54} = \frac{\mathbf{3 \cdot 3 \cdot 5}}{\mathbf{2 \cdot 3 \cdot 3 \cdot 3}} = \mathbf{\frac{5}{6}}$

8. $\frac{1}{4}$ $\quad 4\overline{)1.00}$ = **0.25**

9. $12\% = \frac{12}{100} = \mathbf{0.12} = \mathbf{\frac{3}{25}}$

10. Multiples of 6: 6, 12, 18, 24, 30, …
Multiples of 8: 8, 16, 24, 32, 40, …
Multiples of 12: 12, 24, 36, 48, 60, …
The LCM of 6, 8, and 12 is **24.**

11. $6 - (3.7 + 0.48) = 6 - 4.18 =$ **1.82**

12.
$$\begin{array}{r} 1.2 \\ \times\ 0.3 \\ \hline \mathbf{0.36} \end{array}$$

13. $0.4\overline{)5.00}$ = **12.5**
$$\begin{array}{r} 4 \\ \hline 10 \\ 8 \\ \hline 20 \\ 20 \\ \hline 0 \end{array}$$

14. $14.5 \div 100 =$ **0.145**

15.
$$\begin{array}{r} \frac{1}{2} = \frac{6}{12} \\ \frac{2}{3} = \frac{8}{12} \\ +\ \frac{3}{4} = \frac{9}{12} \\ \hline \frac{23}{12} = \mathbf{1\frac{11}{12}} \end{array}$$

16.
$$\begin{array}{r} 3\frac{3}{5} = 3\frac{6}{10} \\ +\ 6\frac{1}{2} = 6\frac{5}{10} \\ \hline 9\frac{11}{10} = \mathbf{10\frac{1}{10}} \end{array}$$

17.
$$\begin{array}{r} 6\frac{1}{4} = 5\frac{10}{8} \\ -\ 3\frac{7}{8} = 3\frac{7}{8} \\ \hline \mathbf{2\frac{3}{8}} \end{array}$$

18. $3\frac{1}{3} \times 1\frac{1}{5} = \frac{10}{3} \times \frac{6}{5} = \frac{60}{15} = \mathbf{4}$

19. $5 \div 1\frac{2}{3} = \frac{5}{1} \times \frac{3}{5} = \mathbf{3}$

20. $12.6 - w = 8.3$
$$\begin{array}{r} 12.6 \\ -\ 8.3 \\ \hline \mathbf{4.3} \end{array}$$

TEST 15, FORM B

1. 9 cm · 8 cm = **72 sq. cm**

2.
$$\begin{array}{r} \$5.90 \\ \times\ 0.08 \\ \hline \$0.4720 \end{array} \qquad \begin{array}{r} \$5.90 \\ +\ \$0.47 \\ \hline \$6.37 \end{array} \qquad \begin{array}{r} \$10.00 \\ -\ \$6.37 \\ \hline \mathbf{\$3.63} \end{array}$$

3. $\frac{3}{5}$ ⓢ$<$ $\frac{13}{20}$ because $\frac{12}{20} < \frac{13}{20}$

4. **Point *H***

5. **(2, –3)**

6. An octagon has **8 sides.**

7. $\frac{54}{81} = \frac{\mathbf{2 \cdot 3 \cdot 3 \cdot 3}}{\mathbf{3 \cdot 3 \cdot 3 \cdot 3}} = \mathbf{\frac{2}{3}}$

8. $\frac{4}{5}$ $\quad 5\overline{)4.0}$ = **0.8**

9. $15\% = \frac{15}{100} = \mathbf{0.15} = \mathbf{\frac{3}{20}}$

10. Multiples of 4: 4, 8, 12, 16, 20, 24, …
Multiples of 8: 8, 16, 24, 32, 40, 48, …
Multiples of 12: 12, 24, 36, 48, 60, 72, …
The LCM of 4, 8, and 12 is **24.**

11. 8 – (4.3 + 0.75) = 8 – 5.05 = **2.95**

12.
$$\begin{array}{r} 0.45 \\ \times\ 0.2 \\ \hline 0.090 \end{array} = \mathbf{0.09}$$

13. $0.5\overline{)4.0}$ = **8**

14. 12.5 ÷ 100 = **0.125**

15.
$$\begin{array}{r} \frac{1}{2} = \frac{4}{8} \\ \frac{3}{4} = \frac{6}{8} \\ +\ \frac{3}{8} = \frac{3}{8} \\ \hline \frac{13}{8} = \mathbf{1\frac{5}{8}} \end{array}$$

16.
$$\begin{array}{r} 7\frac{1}{2} = 7\frac{3}{6} \\ +\ 3\frac{5}{6} = 3\frac{5}{6} \\ \hline 10\frac{8}{6} = \mathbf{11\frac{1}{3}} \end{array}$$

17.
$$\begin{array}{r} 5\frac{2}{3} = 4\frac{20}{12} \\ -\ 1\frac{3}{4} = 1\frac{9}{12} \\ \hline \mathbf{3\frac{11}{12}} \end{array}$$

18. $4\frac{1}{2} \times 2\frac{2}{3} = \frac{9}{2} \times \frac{8}{3} = \frac{72}{6} = \mathbf{12}$

19. $4 \div 1\frac{1}{3} = \frac{4}{1} \times \frac{3}{4} = \frac{12}{4} = \mathbf{3}$

20. $14.5 - m = 3.7$
$$\begin{array}{r} 14.5 \\ -\ 3.7 \\ \hline \mathbf{10.8} \end{array}$$

TEST 16, FORM A

1. $\mathbf{\frac{1}{4}}$

2. $\frac{42}{50} = \frac{84}{100} = \mathbf{84\%}$

3. $\dfrac{6\text{ cm} \cdot 8\text{ cm}}{2} = \mathbf{24\ sq.\ cm}$

4. Perimeter $= 3 + 2 + 3 + 2 =$ **10 units**

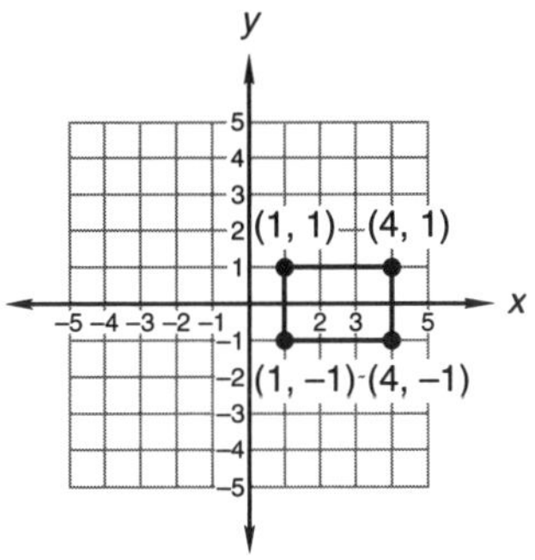

5. Area $= 3 \times 2 =$ **6 sq. units**

6. $\dfrac{4}{5}$ ⓐ> 0.6 because $0.8 > 0.6$

7. $\dfrac{1}{3} \times \dfrac{12}{1} = 4$ $\quad 12 - 4 =$ **8 eggs**

8. **False**

9. $60\% = \dfrac{60}{100} = \dfrac{3}{5}$ $\quad \dfrac{3}{5} \times \dfrac{30}{1} =$ **18 girls**

10. $0.5\overline{)5.0}$ = **10**

11. $4.32 + 5.0 + 6.7 =$ **16.02**

12. $2.8 \div \sqrt{64} = 2.8 \div 8$ $\quad 8\overline{)2.80}$ = **0.35**

$$\begin{array}{r} 2.80 \\ -\,2\,4 \\ \hline 40 \\ -\,40 \\ \hline 0 \end{array}$$

13. $3.6 \div 10^2 = 3.6 \div 100 =$ **0.036**

14. $\dfrac{2}{3} = \dfrac{4}{6}$

$-\,\dfrac{1}{6} = \dfrac{1}{6}$

$\dfrac{3}{6} = \mathbf{\dfrac{1}{2}}$

15. $4\dfrac{1}{2} = 4\dfrac{2}{4}$

$+\,2\dfrac{3}{4} = 2\dfrac{3}{4}$

$6\dfrac{5}{4} - \mathbf{7\dfrac{1}{4}}$

16. $4\dfrac{1}{4} = 3\dfrac{5}{4}$

$-\,3\dfrac{3}{4} = 3\dfrac{3}{4}$

$\dfrac{2}{4} = \mathbf{\dfrac{1}{2}}$

17. $2\dfrac{1}{2} \times 1\dfrac{1}{5} = \dfrac{5}{2} \times \dfrac{6}{5} = \mathbf{3}$

18. $\dfrac{9}{25} \times \dfrac{4}{4} = \mathbf{\dfrac{36}{100}}$

19. $2 \div 1\dfrac{2}{3} = \dfrac{2}{1} \times \dfrac{3}{5} = \dfrac{6}{5} = \mathbf{1\dfrac{1}{5}}$

20. $0.5678 \approx$ **0.57**

Test 16, Form B

1. **2 quarts**

2. $\dfrac{43}{50} = \dfrac{86}{100} =$ **86%**

3. $\dfrac{3\text{ cm} \cdot 4\text{ cm}}{2} = \mathbf{6\ sq.\ cm}$

4. Perimeter = 3 + 3 + 3 + 3 = **12 units**

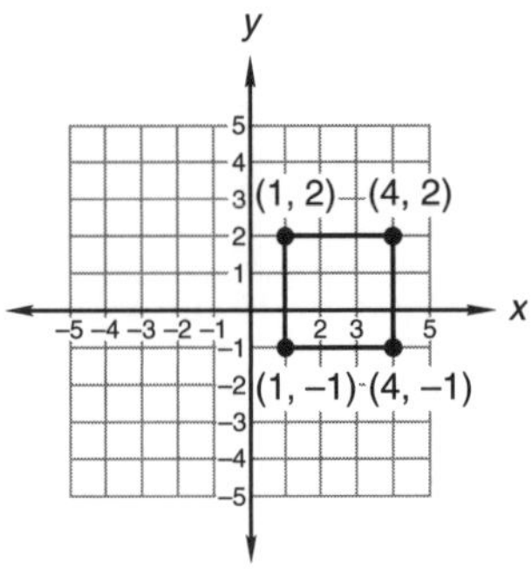

5. Area = 3 × 3 = **9 sq. units**

6. $\frac{1}{4}$ ⓐ 0.3 because 0.25 < 0.3

7. $\frac{1}{6} \times \frac{12}{1} = 2$ cracked eggs

12 − 2 = **10 eggs**

8. **True**

9. $40\% = \frac{40}{100} = \frac{2}{5}$

$\frac{2}{5} \times 60$ minutes = **24 minutes**

10. $0.4\overline{)4.0}$ = **10**

11.
```
  12.0
   4.87
+  6.3
 23.17
```

12. $3.4 \div \sqrt{25} = 3.4 \div 5$
```
   0.68
5)3.40
  3 0
    40
    40
     0
```

13. $5.2 \div 10^2 = 5.2 \div 100 =$ **0.052**

14. $\frac{5}{6} = \frac{5}{6}$

$-\ \frac{1}{3} = \frac{2}{6}$

$\frac{3}{6} = \mathbf{\frac{1}{2}}$

15. $6\frac{5}{8} = 6\frac{5}{8}$

$+\ 1\frac{1}{2} = 1\frac{4}{8}$

$7\frac{9}{8} = \mathbf{8\frac{1}{8}}$

16. $6\frac{1}{2} = 5\frac{12}{8}$

$-\ 2\frac{7}{8} = 2\frac{7}{8}$

$\mathbf{3\frac{5}{8}}$

17. $1\frac{1}{4} \times 2\frac{2}{5} = \frac{5}{4} \times \frac{12}{5} = \mathbf{3}$

18. $\frac{7}{25} \times \frac{4}{4} = \mathbf{\frac{28}{100}}$

19. $2 \div 2\frac{1}{3} = \frac{2}{1} \times \frac{3}{7} = \mathbf{\frac{6}{7}}$

20. 5.4638 ≈ **5.5**

Test 17, Form A

1. $80\% = \frac{80}{100} = \mathbf{\frac{4}{5}}$ = **0.80** = **0.8**

2. $\frac{16\text{ cm} \cdot 6\text{ cm}}{2} = \frac{96\text{ cm}^2}{2} = \mathbf{48\text{ cm}^2}$

3. $\frac{3}{5} = \frac{60}{100} =$ **60%**

4. 3.14 × 100 cm = **314 cm**

5. Perimeter = 10 mm + 12 mm + 10 mm + 12 mm
= **44 mm**

6. Area = 10 mm × 9 mm = **90 sq. mm**

7. Opposite angles in a parallelogram are equal.
$m\angle A = m\angle C =$ **50°**

8. $\frac{\text{pigs}}{\text{goats}} = \frac{12}{8} = \mathbf{\frac{3}{2}}$

9.

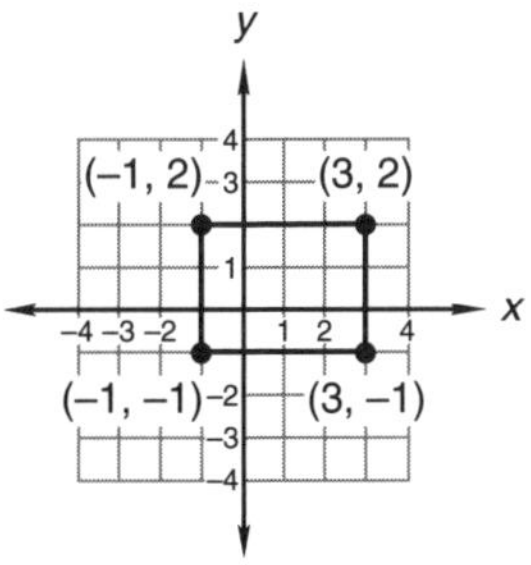

The fourth vertex is at **(–1, 2).**

10. $\frac{3}{4} \times \frac{2000}{1} = \mathbf{1500\ tickets}$

11. 3 ft = 36 in. 36 in. – 10 in. = **26 in.**

12. $3 \times 4 + 2 \times 5 = 12 + 10 = \mathbf{22}$

13. (12 in.)(6 in.)(5 in.) = $\mathbf{360\ in.^3}$

14. $\frac{6}{9} = \frac{10}{w}$
$6w = 9 \cdot 10$
$w = \frac{90}{6} = \mathbf{15}$

15. $10 - (0.1 + 1) = 10 - 1.1 = \mathbf{8.9}$

16. $3n = 0.45$
$n = \frac{0.45}{3}$

$$\begin{array}{r} \mathbf{0.15} \\ 3\overline{)0.45} \\ \underline{3} \\ 15 \\ \underline{15} \\ 0 \end{array}$$

17.
$$\begin{array}{rl} 3\frac{3}{4} = & 3\frac{6}{8} \\ + 1\frac{5}{8} = & 1\frac{5}{8} \\ \hline & 4\frac{11}{8} = \mathbf{5\frac{3}{8}} \end{array}$$

18.
$$\begin{array}{rl} 2\frac{1}{2} = & 1\frac{6}{4} \\ - \frac{3}{4} = & \frac{3}{4} \\ \hline & \mathbf{1\frac{3}{4}} \end{array}$$

19. $2\frac{1}{4} \times 2 = \frac{9}{4} \times \frac{2}{1} = \frac{9}{2} = \mathbf{4\frac{1}{2}}$

20. $5 \div 3\frac{3}{4} = \frac{5}{1} \times \frac{4}{15} = \frac{4}{3} = \mathbf{1\frac{1}{3}}$

Test 17, Form B

1. $60\% = \frac{60}{100} = \mathbf{\frac{3}{5}} = \mathbf{0.60} = \mathbf{0.6}$

2. $\frac{12\text{ cm} \cdot 8\text{ cm}}{2} = \frac{96\text{ cm}^2}{2} = \mathbf{48\ cm^2}$

3. $\frac{4}{5} = \frac{80}{100} = \mathbf{80\%}$

4. $3.14 \times 200\text{ cm} = \mathbf{628\ cm}$

5. Perimeter = 10 cm + 8 cm + 10 cm + 8 cm
= **36 cm**

6. Area = 10 cm × 6 cm = **60 sq. cm**

7. Adjacent angles in a parallelogram are supplementary.
m∠*C* = 180° m∠*B* = 180° – 130° = **50°**

8. $\frac{\text{children}}{\text{adults}} = \frac{80}{60} = \mathbf{\frac{4}{3}}$

9.

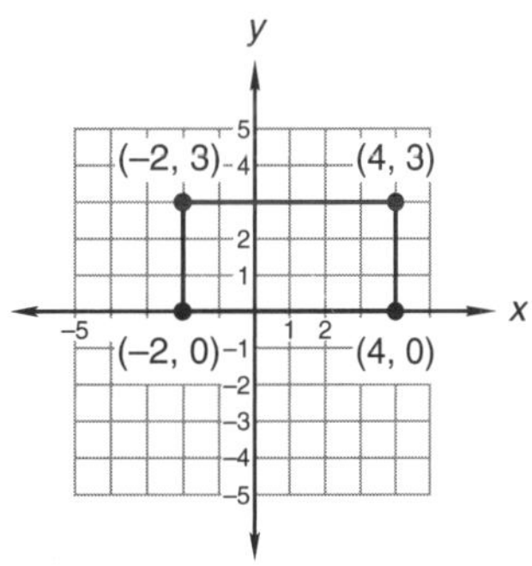

The fourth vertex is at **(–2, 3).**

10. $\frac{4}{5} \times \frac{20}{1} = \mathbf{16\ questions}$

11. 3 ft = 36 in. 36 in. – 6 in. = **30 in.**

12. $3 \times 4 - 2 \times 5 = 12 - 10 = \mathbf{2}$

13. (10 in.)(4 in.)(5 in.) = $\mathbf{200\ in.^3}$

14. $\dfrac{9}{m} = \dfrac{6}{10}$

$6m = 9 \cdot 10$

$m = \dfrac{90}{6} = \mathbf{15}$

15. $2 - (0.2 + 0.02) = 2 - 0.22 = \mathbf{1.78}$

16. $4n = 0.56$

$n = \dfrac{0.56}{4}$

$$\begin{array}{r} \mathbf{0.14} \\ 4\overline{)0.56} \\ \underline{4} \\ 16 \\ \underline{16} \\ 0 \end{array}$$

17.
$$\begin{array}{rcl} 1\frac{3}{4} & = & 1\frac{3}{4} \\ + \ 1\frac{1}{2} & = & 1\frac{2}{4} \\ \hline & & 2\frac{5}{4} = \mathbf{3\frac{1}{4}} \end{array}$$

18.
$$\begin{array}{rcl} 4\frac{1}{2} & = & 3\frac{6}{4} \\ - \ 1\frac{3}{4} & = & 1\frac{3}{4} \\ \hline & & \mathbf{2\frac{3}{4}} \end{array}$$

19. $5 \times 1\frac{3}{5} = \frac{5}{1} \times \frac{8}{5} = \mathbf{8}$

20. $5 \div 6\frac{2}{3} = \frac{5}{1} \times \frac{3}{20} = \mathbf{\frac{3}{4}}$

TEST 18, FORM A

1. $\frac{3}{4} \times \frac{200}{1} = 150$ seeds sprouted

$200 - 150 =$ **50 seeds** did not sprout

2. $\dfrac{\text{boys}}{\text{girls}} = \dfrac{3}{2} = \dfrac{12}{g}$

$3g = 24$

$g =$ **8 girls**

3. Perimeter = 6 cm + 8 cm + 10 cm = **24 cm**

4. Area = $\dfrac{6 \text{ cm} \times 8 \text{ cm}}{2}$ = **24 sq. cm**

5.
$$\begin{array}{r} 5.00 \\ -\ 3.21 \\ \hline \mathbf{1.79} \end{array}$$

6.
$$\begin{array}{r} 2.7 \\ \times\ 0.03 \\ \hline \mathbf{0.081} \end{array}$$

7. $0.27 \div \sqrt{36} = 0.27 \div 6$

$$\begin{array}{r} \mathbf{0.045} \\ 6\overline{)0.270} \\ \underline{24} \\ 30 \\ \underline{30} \\ 0 \end{array}$$

8. $$\begin{array}{r} \mathbf{30} \\ 0.9\overline{)27.0} \end{array}$$

9. $4^2 = 16$ and $5^2 = 25$

C. 4 and 5

10.
$$\begin{array}{rcl} \frac{5}{6} & = & \frac{5}{6} \\ + \ \frac{1}{2} & = & \frac{3}{6} \\ \hline & & \frac{8}{6} = \mathbf{1\frac{1}{3}} \end{array}$$

11.
$$\begin{array}{rcl} 3\frac{1}{3} & = & 2\frac{8}{6} \\ - \ 2\frac{1}{2} & = & 2\frac{3}{6} \\ \hline & & \mathbf{\frac{5}{6}} \end{array}$$

12. $3\frac{3}{4} \times 1\frac{3}{5} = \frac{15}{4} \times \frac{8}{5} = \mathbf{6}$

13. $5 \div 1\frac{1}{4} = \frac{5}{1} \times \frac{4}{5} = \mathbf{4}$

14. $5y = 9$

$y = \frac{9}{5} = \mathbf{1\frac{4}{5}}$

15. 3.14 × 20 cm = **62.8 cm**

16. 3.14 × 10 cm × 10 cm = **314 cm²**

17. $75\% = \frac{75}{100} = \mathbf{\frac{3}{4}} = \mathbf{0.75}$

18. (10 cm)(6 cm)(5 cm) = **300 cm³**

19. **6 faces**

20. **(–2, –2), (8, –2), or (2, 6)**

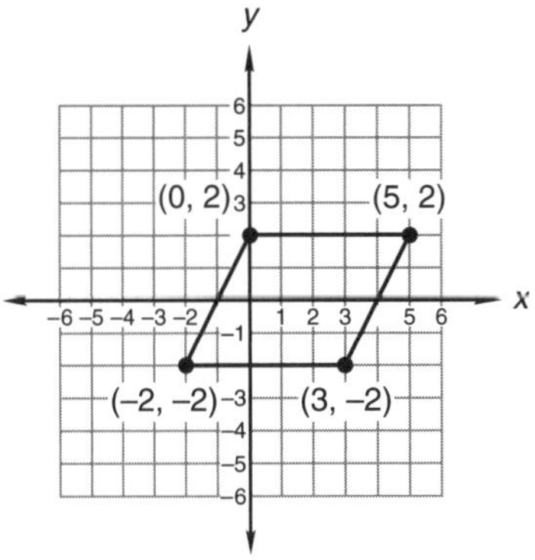

Test 18, Form B

1. $\frac{4}{5} \times \frac{200}{1} = 160$ seeds sprouted

200 – 160 = **40 seeds** did not sprout

2. $\frac{\text{boys}}{\text{girls}} = \frac{3}{2} = \frac{b}{12}$

$2b = 36$

$b =$ **18 boys**

3. Perimeter = 5 cm + 3 cm + 4 cm = **12 cm**

4. Area = $\frac{3 \text{ cm} \times 4 \text{ cm}}{2}$ = **6 sq. cm**

5.
```
  6.0
– 4.3
  1.7
```
(answer **1.7**)

6.
```
   4.3
× 0.02
 0.086
```
(answer **0.086**)

7. $0.45 \div \sqrt{36} = 0.45 \div 6$

```
   0.075
6)0.450
    42
     30
     30
      0
```
(answer **0.075**)

8.
```
    50
0.9)45.0
```
(answer **50**)

9. $5^2 = 25$ and $6^2 = 36$

C. 5 and 6

10. $\frac{2}{3} = \frac{4}{6}$

$+ \frac{5}{6} = \frac{5}{6}$

$\frac{9}{6} = \mathbf{1\frac{1}{2}}$

11. $3\frac{1}{4} = 2\frac{5}{4}$

$- 2\frac{1}{2} = 2\frac{2}{4}$

$\mathbf{\frac{3}{4}}$

12. $1\frac{1}{5} \times 1\frac{2}{3} = \frac{6}{5} \times \frac{5}{3} = \mathbf{2}$

13. $5 \div 1\frac{2}{3} = \frac{5}{1} \times \frac{3}{5} = \mathbf{3}$

14. $5y = 8$

$y = \frac{8}{5} = \mathbf{1\frac{3}{5}}$

15. 3.14 × 20 in. = **62.8 in.**

16. 3.14 × 10 in. × 10 in. = **314 in.²**

17. $65\% = \frac{65}{100} = \mathbf{\frac{13}{20}} = \mathbf{0.65}$

18. (10 cm)(10 cm)(10 cm) = **1000 cm³**

19. **12 edges**

20. **(1, 0), (–5, 0), or (5, 6)**

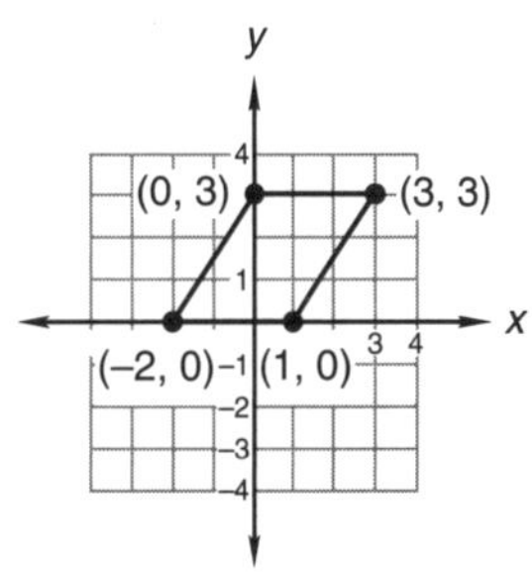

Test 19, Form A

1. $\frac{\text{favorable}}{\text{possible}} = \frac{3}{6} = \mathbf{\frac{1}{2}}$

2. 85, 85, 85, **90,** 90, 100, 100

3. $\frac{2}{8} = \frac{1}{4} = \frac{25}{100} = \mathbf{25\%}$

4. **8 edges**

5. $\text{Area} = \frac{9 \text{ cm} \times 12 \text{ cm}}{2} = \mathbf{54 \text{ sq. cm}}$

6. **C. right triangle**

7. $10 + 5 \times 4 - 2^2 = 10 + 20 - 4 = \mathbf{26}$

8. $\left(1\frac{1}{2}\right)^2 = \frac{3}{2} \cdot \frac{3}{2} = \frac{9}{4} = \mathbf{2\frac{1}{4}}$

9. $\frac{4}{6} = \frac{10}{w}$
$4w = 60$
$w = \frac{60}{4} = \mathbf{15}$

10. $4^2 - \sqrt{16} = 16 - 4 = \mathbf{12}$

11. $\frac{(15)(16)}{20} = \frac{240}{20} = \mathbf{12}$

12. $\frac{5}{8} \times \frac{56}{1} = \mathbf{35}$

13. $6 - (0.37 + 1.2) = 6 - 1.57 = \mathbf{4.43}$

14. $0.5\overline{)4.0}$ = **8**

15. $6\frac{2}{3} = 6\frac{4}{6}$
$3 = 3$
$+ 4\frac{1}{2} = 4\frac{3}{6}$
$13\frac{7}{6} = \mathbf{14\frac{1}{6}}$

16. $3\frac{1}{3} = 2\frac{16}{12}$
$- 2\frac{3}{4} = 2\frac{9}{12}$
$\mathbf{\frac{7}{12}}$

17. $1\frac{4}{5} \div 3 = \frac{9}{5} \times \frac{1}{3} = \mathbf{\frac{3}{5}}$

18. $\frac{1}{3} \times 100\% = \frac{100\%}{3} = \mathbf{33\frac{1}{3}\%}$

19. $3\frac{3}{4} = \mathbf{3.75}$
$3.75 + 1.5 = \mathbf{5.25}$

20. $\text{Perimeter} = 4 + 6 + 4 + 6 = \mathbf{20 \text{ units}}$

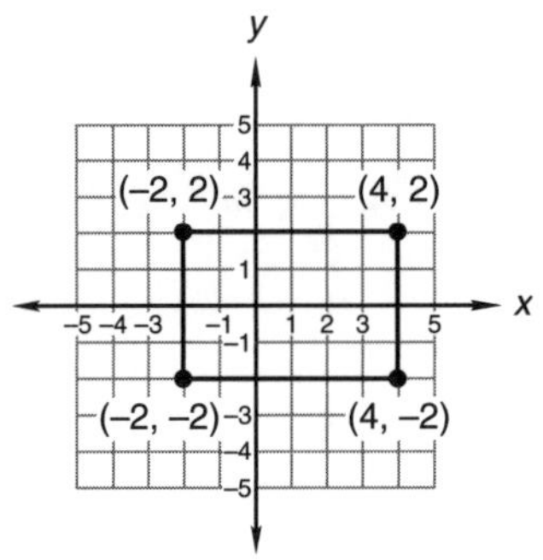

Test 19, Form B

1. $\frac{\text{favorable}}{\text{possible}} = \frac{3}{6} = \mathbf{\frac{1}{2}}$

2. 17, 18, 18, **19,** 19, 20, 20

3. $\frac{2}{5} = \frac{40}{100} = \mathbf{40\%}$

4. **5 faces**

5. $\text{Area} = \frac{5 \text{ cm} \times 12 \text{ cm}}{2} = \mathbf{30 \text{ sq. cm}}$

6. **D. right triangle**

7. $8 + 4 \times 2 - 3^2 = 8 + 8 - 9 = \mathbf{7}$

8. $\left(2\frac{1}{2}\right)^2 = \frac{5}{2} \cdot \frac{5}{2} = \frac{25}{4} = \mathbf{6\frac{1}{4}}$

9. $\frac{6}{9} = \frac{8}{n}$
$6n = 72$
$n = \frac{72}{6} = \mathbf{12}$

10. $3^3 - \sqrt{9} = 27 - 3 = \mathbf{24}$

11. $\frac{(25)(18)}{15} = \frac{450}{15} = \mathbf{30}$

12. $\frac{3}{5} \times \frac{60}{1} = \mathbf{36}$

13. $6 + (1.2 - 0.37) = 6 + 0.83 = \mathbf{6.83}$

14. $0.5\overline{)3.0}$ = **6**

15.
$3\frac{3}{8} = 3\frac{3}{8}$
$4 = 4$
$+ 2\frac{3}{4} = 2\frac{6}{8}$
$9\frac{9}{8} = \mathbf{10\frac{1}{8}}$

16.
$3\frac{2}{3} = 2\frac{20}{12}$
$- 2\frac{3}{4} = 2\frac{9}{12}$
$\mathbf{\frac{11}{12}}$

17. $5\frac{1}{3} \div 4 = \frac{16}{3} \times \frac{1}{4} = \frac{4}{3} = \mathbf{1\frac{1}{3}}$

18. $\frac{2}{3} \times 100\% = \frac{200\%}{3} = \mathbf{66\frac{2}{3}\%}$

19. $3\frac{3}{4} = \mathbf{3.75}$
$3.75 + 2.5 = \mathbf{6.25}$

20. Perimeter = 5 + 4 + 5 + 4 = **18 units**

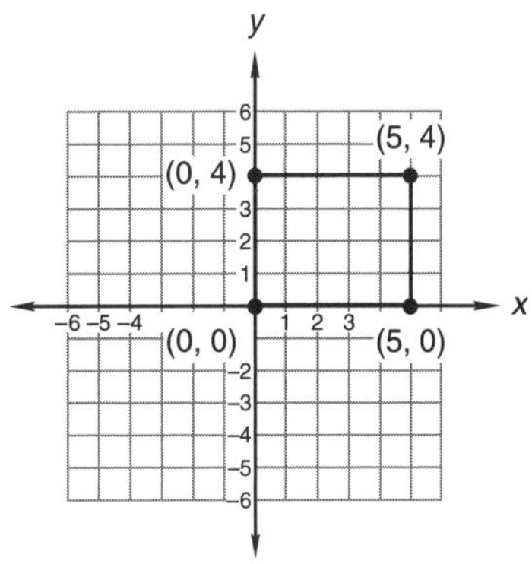

TEST 20, FORM A

1. $\frac{3}{4} = \frac{12}{B}$
$3B = 48$
$B = \frac{48}{3} =$ **16 B's**

2. $\frac{1}{6} \times \frac{1}{6} = \mathbf{\frac{1}{36}}$

3. (a) $0.6 = \frac{6}{10} = \mathbf{\frac{3}{5}}$

(b) $0.6 = \frac{6}{10} = \frac{60}{100} = \mathbf{60\%}$

4. (a) $5\% = \frac{5}{100} = \mathbf{\frac{1}{20}}$

(b) $5\% = \frac{5}{100} = \mathbf{0.05}$

5. $\frac{3}{5} \times 30 = 18$ students finished
$30 - 18 =$ **12 students** did not finish

6. Area = 12 mm × 15 mm = **180 sq. mm**

7. Perimeter = 13 mm + 15 mm + 13 mm + 15 mm
= **56 mm**

8. $m\angle 6 = m\angle 2 = \mathbf{80°}$

9. **360°**

10. $\sqrt{900} = \mathbf{30}$

11. $-3 + -5 = \mathbf{-8}$

12. $-5 - (-7) = -5 + 7 = \mathbf{2}$

13. $5 \text{ cm} \times 4 \text{ cm} \times 3 \text{ cm} = \mathbf{60 \text{ cm}^3}$

14. $12 - 8 = \mathbf{4}$

15. $2^3 - \sqrt{9} + 3 \times 2 = 8 - 3 + 6 = \mathbf{11}$

16. $6.25 - 4\frac{1}{2} = 6.25 - 4.5 = \mathbf{1.75}$

17. $\frac{3}{20} = \frac{m}{100}$
$20m = 300$
$m = \mathbf{15}$

18. $y + 3.6 = 5$
$y = 5 - 3.6 = \mathbf{1.4}$

19. $6°\text{F} + 12°\text{F} = \mathbf{18°F}$

20. $3.14 \times 6 \text{ in.} \times 6 \text{ in.} = \mathbf{113.04 \text{ in.}^2}$

TEST 20, FORM B

1. $\frac{4}{3} = \frac{12}{B}$
$4B = 36$
$B = \frac{36}{4} = $ **9 B's**

2. $\frac{1}{6} \times \frac{1}{6} = \mathbf{\frac{1}{36}}$

3. (a) $0.4 = \frac{4}{10} = \mathbf{\frac{2}{5}}$

(b) $0.4 = \frac{4}{10} = \frac{40}{100} = \mathbf{40\%}$

4. (a) $4\% = \frac{4}{100} = \mathbf{\frac{1}{25}}$

(b) $4\% = \frac{4}{100} = \mathbf{0.04}$

5. $\frac{2}{3} \times 30 = 20$ students finished
$30 - 20 = $ **10 students** did not finish

6. Area $= 6 \text{ cm} \times 8 \text{ cm} = $ **48 sq. cm**

7. Perimeter $= 10 \text{ cm} + 6 \text{ cm} + 10 \text{ cm} + 6 \text{ cm}$
$= $ **32 cm**

8. $m\angle 5 = m\angle 1 = \mathbf{105°}$

9. **360°**

10. $\sqrt{400} = \mathbf{20}$

11. $-2 + -5 = \mathbf{-7}$

12. $-3 - (-7) = -3 + 7 = \mathbf{4}$

13. $3 \text{ cm} \times 3 \text{ cm} \times 3 \text{ cm} = \mathbf{27 \text{ cm}^3}$

14. $8 - 6 = \mathbf{2}$

15. $2^3 + \sqrt{9} - 3 \times 2 = 8 + 3 - 6 = \mathbf{5}$

16.
$$\begin{array}{rr} 6\frac{1}{2} = & 6.5 \\ +\ 4.25 = & 4.25 \\ \hline & \mathbf{10.75} \end{array}$$

17. $\frac{3}{25} = \frac{m}{100}$
$25m = 300$
$m = \mathbf{12}$

18. $y + 2.7 = 5$
$y = 5 - 2.7 = \mathbf{2.3}$

19. $5°\text{F} + 8°\text{F} = \mathbf{13°F}$

20. $3.14 \times 8 \text{ in.} \times 8 \text{ in.} = \mathbf{200.96 \text{ in.}^2}$

Test 21, Form A

1. $\$0.1625$ rounds to **$0.16**

$$\begin{array}{r} 0.1625 \\ 8\overline{)\$1.3000} \\ \underline{8} \\ 50 \\ \underline{48} \\ 20 \\ \underline{16} \\ 40 \\ \underline{40} \\ 0 \end{array}$$

2.

	Ratio	Actual Count
Earned an A	40%	10
Did not earn an A	60%	*n*
Total	100%	*s*

$$\frac{40}{100} = \frac{10}{s}$$
$$40s = 1000$$
$$s = \mathbf{25\ students}$$

3. **(1, –2)**

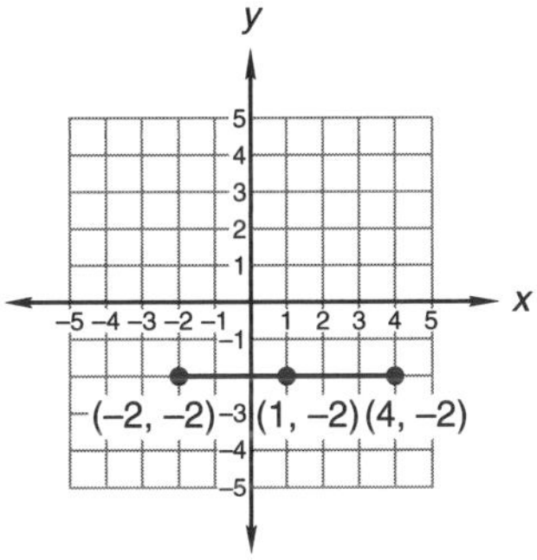

4. $10^2 - \sqrt{81} - 3 \times 10 = 100 - 9 - 30 = \mathbf{61}$

5. 4 in. × 2 in. × 3 in. = 24 cu. in.
24 sugar cubes would be needed.

6. 7 cm – 3 cm = 4 cm
5 cm – 3 cm = 2 cm
Perimeter = 3 cm + 5 cm + 7 cm + 3 cm + 4 cm + 2 cm = **24 cm**

7. 3.14 × 20 cm = **62.8 cm**

8. $2 \text{ tons} \times \frac{2000 \text{ pounds}}{1 \text{ ton}} = \mathbf{4000\ pounds}$

9. $$\frac{6}{10} = \frac{a}{25}$$
$$10a = 6 \cdot 25$$
$$a = \frac{150}{10}$$
$$a = \mathbf{15}$$

10.

	Ratio	Actual Count
Boys	5	*b*
Girls	4	*g*
Total	9	27

$$\frac{5}{9} = \frac{b}{27}$$
$$9b = 135$$
$$b = \frac{135}{9} = \mathbf{15\ boys}$$

11. (a) $1\frac{3}{5} = \frac{8}{5}$ $\qquad 5\overline{)8.0}$ with quotient **1.6**

(b) $1\frac{3}{5} = 1.6 \times 100\% = \mathbf{160\%}$

12. (a) $0.9 = \mathbf{\frac{9}{10}}$

(b) $0.9 \times 100\% = \mathbf{90\%}$

13. (a) $10\% = \frac{10}{100} = \mathbf{\frac{1}{10}}$

(b) $10\% = 0.10 = \mathbf{0.1}$

14. $6.3 - (1.25 + 2) = 6.3 - 3.25 = \mathbf{3.05}$

15. $-3 - 2 + 4 = \mathbf{-1}$

16. $4 \div (0.2 \times 2) = 4 \div 0.4 = \mathbf{10}$

17. $$\begin{array}{r} 6\frac{2}{3} = 6\frac{8}{12} \\ + 3\frac{3}{4} = 3\frac{9}{12} \\ \hline 9\frac{17}{12} = \mathbf{10\frac{5}{12}} \end{array}$$

18. $6\frac{1}{4} \times 1\frac{3}{5} = \frac{25}{4} \times \frac{8}{5} = \mathbf{10}$

19. $2\frac{1}{3} \div 1\frac{5}{9} = \frac{7}{3} \times \frac{9}{14} = \frac{3}{2} = \mathbf{1\frac{1}{2}}$

20. $\text{Area} = \frac{6 \text{ cm} \cdot 8 \text{ cm}}{2} = \mathbf{24 \text{ sq. cm}}$

Test 21, Form B

1. $8\overline{)\$6.5000}$ = $0.8125, rounds to **$0.81**

```
   $0.8125
8)$6.5000
   6 4
     10
      8
      20
      16
       40
       40
        0
```

2.

	Ratio	Actual Count
Earned an A	30%	9
Did not earn an A	70%	n
Total	100%	s

$\frac{30}{100} = \frac{3}{10} = \frac{9}{s}$

$3s = 90$

$s = \mathbf{30 \text{ students}}$

3. **(2, 3)**

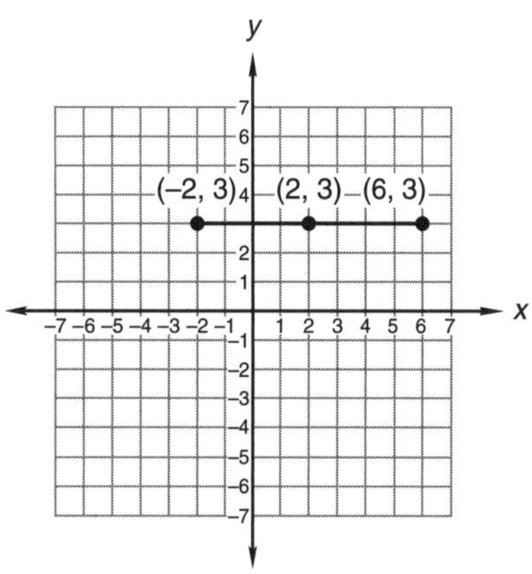

4. $10^2 + \sqrt{9} + 3 \times 10 = 100 + 3 + 30 = \mathbf{133}$

5. $3 \text{ cm} \times 3 \text{ cm} \times 3 \text{ cm} = 27 \text{ cu. cm}$
27 sugar cubes would be needed.

6. $10 \text{ cm} - 5 \text{ cm} = 5 \text{ cm}$
$7 \text{ cm} - 4 \text{ cm} = 3 \text{ cm}$
$\text{Perimeter} = 4 \text{ cm} + 10 \text{ cm} + 7 \text{ cm} + 5 \text{ cm} + 3 \text{ cm} + 5 \text{ cm} = \mathbf{34 \text{ cm}}$

7. $3.14 \times 10 \text{ cm} \times 10 \text{ cm} = \mathbf{314 \text{ cm}^2}$

8. $\frac{1}{2} \text{ ton} \times \frac{2000 \text{ pounds}}{1 \text{ ton}} = \mathbf{1000 \text{ pounds}}$

9. $\frac{8}{12} = \frac{12}{w}$

$8w = 12 \cdot 12$

$8w = 144$

$w = \frac{144}{8}$

$w = \mathbf{18}$

10.

	Ratio	Actual Count
Boys	4	b
Girls	5	g
Total	9	27

$\frac{4}{9} = \frac{b}{27}$

$9b = 108$

$b = \frac{108}{9} = \mathbf{12 \text{ boys}}$

11. (a) $2\frac{1}{5} = \frac{11}{5}$ $\qquad 5\overline{)11.0} = \mathbf{2.2}$

(b) $2\frac{1}{5} = 2.2 \times 100\% = \mathbf{220\%}$

12. (a) $0.7 = \mathbf{\frac{7}{10}}$

(b) $0.7 \times 100\% = \mathbf{70\%}$

13. (a) $5\% = \frac{5}{100} = \mathbf{\frac{1}{20}}$

(b) $5\% = \mathbf{0.05}$

14. $4.2 + (3 - 1.75) = 4.2 + 1.25 = \mathbf{5.45}$

15. $-2 - 3 + 5 = \mathbf{0}$

16. $0.12 \times (0.12 \div 6) = 0.12 \times 0.02 = \mathbf{0.0024}$

17.
$$\begin{array}{r} 1\frac{3}{4} = 1\frac{9}{12} \\ + \ 1\frac{2}{3} = 1\frac{8}{12} \\ \hline 2\frac{17}{12} = \mathbf{3\frac{5}{12}} \end{array}$$

18. $8\frac{1}{3} \times 1\frac{4}{5} = \frac{25}{3} \times \frac{9}{5} = \mathbf{15}$

19. $2\frac{2}{5} \div 1\frac{4}{5} = \frac{12}{5} \times \frac{5}{9} = \frac{4}{3} = \mathbf{1\frac{1}{3}}$

20. $\text{Area} = \frac{6 \text{ cm} \cdot 10 \text{ cm}}{2} = \mathbf{30 \text{ sq. cm}}$

Test 22, Form A

1. $180° - (70° + 70°) = \mathbf{40°}$

2.

3.
$$\begin{array}{r} 4\frac{1}{2} = 3\frac{9}{6} \\ - \ 1\frac{5}{6} = 1\frac{5}{6} \\ \hline 2\frac{4}{6} = \mathbf{2\frac{2}{3}} \end{array}$$

4. $6\frac{2}{3} \div 1\frac{1}{4} = \frac{20}{3} \times \frac{4}{5} = \frac{16}{3} = \mathbf{5\frac{1}{3}}$

5. $3.75 + \frac{3}{8} = 3.75 + 0.375 = \mathbf{4.125}$

6. $\mathbf{-2, 0.2, \frac{1}{2}}$

7.

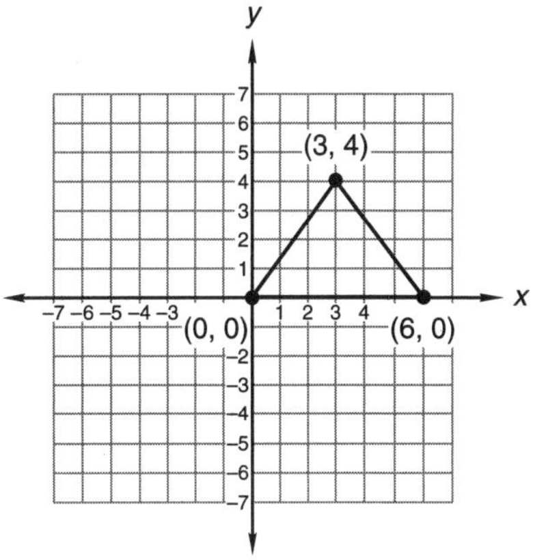

$\text{Area} = \frac{6 \cdot 4}{2} = \mathbf{12 \text{ sq. units}}$

8. $-6 - 4 - 3 = \mathbf{-13}$

9. $\frac{\overset{1}{\cancel{3}}}{\underset{5}{\cancel{10}}} \cdot \frac{\overset{1}{\cancel{2}}}{\underset{3}{\cancel{9}}} = \mathbf{\frac{1}{15}}$

10. $3^3 - 2^3 - 2 \times 3 = 27 - 8 - 6 = \mathbf{13}$

11. (a) $1\frac{1}{5} = \frac{6}{5}$ $\quad 5\overline{)6.0}$ = **1.2**

(b) $1\frac{1}{5} = 1.2 \times 100\% = \mathbf{120\%}$

12. (a) $0.3 = \mathbf{\frac{3}{10}}$

(b) $0.3 \times 100\% = \mathbf{30\%}$

13. (a) $35\% = \frac{35}{100} = \mathbf{\frac{7}{20}}$

(b) $35\% = \mathbf{0.35}$

14. 5 in. × 4 in. × 4 in. = **80 cu. in.**

15. 3.14 × 10 in. × 10 in. = $\mathbf{314 \text{ in.}^2}$

16. $\frac{\text{boy}}{\text{girl}} = \frac{14}{18} = \mathbf{\frac{7}{9}}$

17. $\frac{a}{12} = \frac{20}{30}$
$30a = 12 \cdot 20$
$30a = 240$
$a = \frac{240}{30}$
$a = \mathbf{8}$

18. $3n + 2 = 35$
$3n = 33$
$n = \frac{33}{3} = \mathbf{11}$

19. 10 m − 4 m = 6 m
8 m − 3 m = 5 m
Perimeter = 4 m + 8 m + 10 m + 3 m + 6 m + 5 m = **36 m**

20. Area = (8 m × 4 m) + (6 m × 3 m)
= 32 sq. m + 18 sq. m = **50 sq. m**

Test 22, Form B

1. 180° − (75° + 75°) = **30°**

2.

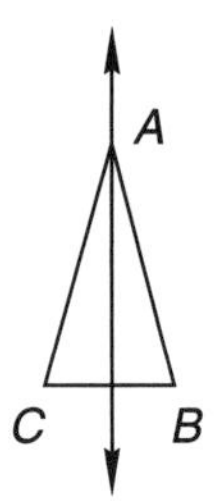

3. $5\frac{1}{3} = 4\frac{8}{6}$
$-\ 3\frac{1}{2} = 3\frac{3}{6}$
$\mathbf{1\frac{5}{6}}$

4. $3\frac{3}{4} \div 1\frac{1}{4} = \frac{15}{4} \times \frac{4}{5} = \mathbf{3}$

5. $4.25 + \frac{3}{8} = 4.25 + 0.375 = \mathbf{4.625}$

6. **−3, 0.35, $\frac{3}{4}$**

7.

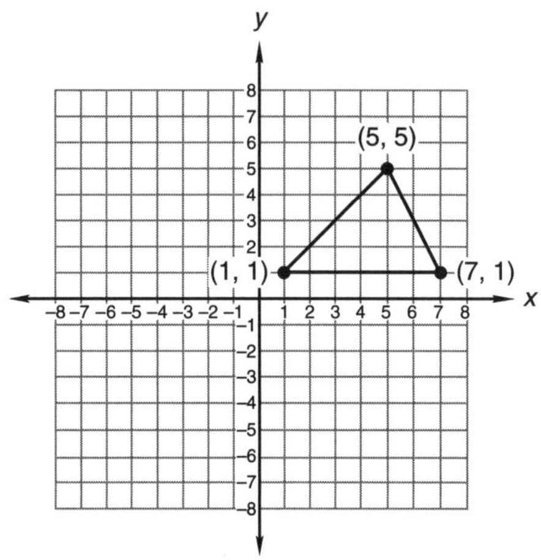

Area = $\frac{6 \cdot 4}{2}$ = **12 sq. units**

8. −7 − 2 − 3 = **−12**

9. $\frac{\overset{1}{\cancel{5}}}{\underset{2}{\cancel{10}}} \cdot \frac{\overset{2}{\cancel{4}}}{9} = \mathbf{\frac{2}{9}}$

10. $3^2 - 2^3 + 2 \times 3 = 9 - 8 + 6 = \mathbf{7}$

11. (a) $1\frac{1}{4} = \frac{5}{4}$ $4\overline{)5.00}$ = **1.25**

(b) $1\frac{1}{4}$ = 1.25 × 100% = **125%**

12. (a) $0.7 = \mathbf{\frac{7}{10}}$

(b) 0.7 × 100% = **70%**

13. (a) $15\% = \frac{15}{100} = \mathbf{\frac{3}{20}}$

(b) 15% = **0.15**

14. 8 × 3 × 4 = **96 blocks**

15. 3.14 × 32 in. = **100.48 in.**

16. $\frac{\text{boy}}{\text{girl}} = \frac{18}{14} = \mathbf{\frac{9}{7}}$

17. $\frac{10}{a} = \frac{25}{30}$
$25a = 10 \cdot 30$
$25a = 300$
$a = \frac{300}{25}$
$a = \mathbf{12}$

18. $4m + 4 = 40$
$4m = 36$
$m = \frac{36}{4} = \mathbf{9}$

19. 15 m − 7 m = 8 m
10 m − 6 m = 4 m
Perimeter = 6 m + 15 m + 10 m + 7 m + 4 m + 8 m = **50 m**

20. Area = (7 m × 10 m) + (6 m × 8 m)
= 70 sq. m + 48 sq. m = **118 sq. m**

Test 23, Form A

1. $\frac{110}{4} = 27 \text{ R } 2$

27, 27, 28, 28

2. $1\frac{1}{2}$ million = 1.5 million = **1,500,000**

3. $81 = 3 \cdot 3 \cdot 3 \cdot 3 = \mathbf{3^4}$

4. $\overset{2}{\cancel{8}}\ \cancel{\text{qt}} \times \frac{1 \text{ gal}}{\underset{1}{\cancel{4}}\ \cancel{\text{qt}}} = \mathbf{2\ gal}$

5. $(5 \times 10^3) + (4 \times 10^1) = 5000 + 40 = \mathbf{5040}$

6. $\frac{3}{6} \cdot \frac{2}{5} = \mathbf{\frac{1}{5}}$

7. 15 cm × 3 = **45 cm**

8. 4 ft + 5 ft + 6 ft = **15 ft**

9. 3.14 × 10 in. = **31.4 in.**

10. 4 m + 5 m = 9 m
12 m − 5 m = 7 m
Perimeter = 12 m + 5 m + 7 m + 4 m + 5 m + 9 m = **42 m**

11. Area = (5 m × 12 m) + (4 m × 5 m)
= 60 sq. m + 20 sq. m = **80 sq. m**

12. (a) $4\overline{)3.00}$ = **0.75**

(b) $\frac{3}{4} = 0.75 \times 100\% = \mathbf{75\%}$

13. (a) $0.02 = \frac{2}{100} = \mathbf{\frac{1}{50}}$

(b) 0.02 × 100% = **2%**

14. (a) $120\% = \frac{120}{100} = \frac{6}{5} = \mathbf{1\frac{1}{5}}$

(b) 120% = 1.20 = **1.2**

15. $\begin{array}{r} 5.00 \\ -\ 0.45 \\ \hline 4.55 \end{array}$ $\quad$ $\begin{array}{r} 4.550 \\ +\ 0.475 \\ \hline \mathbf{5.025} \end{array}$

16. 0.6 ÷ (0.4 × 0.5) = 0.6 ÷ 0.2 = **3**

17. (−4)(−7) = **28**

18. $\frac{-16}{+2} = \mathbf{-8}$

19. Area = $\frac{8 \text{ cm} \times 10 \text{ cm}}{2}$ = **40 sq. cm**

20. 100% − 80% = 20%
$\frac{20}{100} = \frac{5}{x}$
$20x = 500$
$x =$ **25 questions**

TEST 23, FORM B

1. $\frac{85}{3} = 28 \text{ R } 1$

28, 28, 29

2. 1.5 billion = **1,500,000,000**

3. $64 = 2 \cdot 2 \cdot 2 \cdot 2 \cdot 2 \cdot 2 = \mathbf{2^6}$

4. $8 \cancel{\text{qt}} \times \frac{2 \text{ pt}}{1 \cancel{\text{qt}}} = \mathbf{16 \text{ pt}}$

5. $(6 \times 10^2) + (4 \times 10^1) = 600 + 40 = \mathbf{640}$

6. $\frac{3}{6} \cdot \frac{2}{5} = \mathbf{\frac{1}{5}}$

7. Perimeter = 36 in. ÷ 3 = **12 in.**

8. Side length = 20 cm ÷ 4 = 5 cm
Area = 5 cm × 5 cm = **25 sq. cm**

9. Diameter = 2 × 10 in. = 20 in.
Area = 3.14 × 20 in. = **62.8 in.**

10. 10 cm − 6 cm = 4 cm
6 cm − 4 cm = 2 cm
Perimeter = 6 cm + 4 cm + 10 cm + 6 cm + 4 cm + 2 cm = **32 cm**

11. Area = (10 cm × 4 cm) + (2 cm × 4 cm)
= 40 sq. cm + 8 sq. cm = **48 sq. cm**

12. (a) $4\overline{)1.00}$ = **0.25**

(b) $\frac{1}{4} = 0.25 \times 100\% = \mathbf{25\%}$

13. (a) $0.04 = \frac{4}{100} = \mathbf{\frac{1}{25}}$

(b) 0.04 × 100% = **4%**

14. (a) $150\% = \frac{150}{100} = \frac{3}{2} = \mathbf{1\frac{1}{2}}$

(b) 150% = 1.50 = **1.5**

15. 6.23 + (4 − 0.5) = 6.23 + 3.5 = **9.73**

16. 6 ÷ (0.4 × 0.5)
6 ÷ 0.2 = **30**

17. (−9)(−4) = **36**

18. $\frac{-18}{+3} = \mathbf{-6}$

19. $\text{Area} = \frac{8 \text{ mm} \times 12 \text{ mm}}{2} = \mathbf{48 \text{ sq. mm}}$

20. 100% − 80% = 20%
$\frac{20}{100} = \frac{6}{x}$
$20x = 600$
$x = \mathbf{30 \text{ questions}}$

FINAL EXAM

1. $25\overline{)7610}$ = $304\frac{10}{25}$

```
   304 10/25
25)7610
   75
    11
     0
    110
    100
     10
```

$304\frac{10}{25} = 304\frac{2}{5}$

C. $\mathbf{304\frac{2}{5}}$

2.
```
   ¹
   8.75
   6.00
 + 4.50
  19.25
```

D. 19.25

3.
```
    7 1
   1.8̸0
 − 0.25
   1.55
```

A. 1.55

4.
```
    0.15
  ×  4.2
     030
    060
   0.630
```

A. 0.63

5. $$\begin{array}{r} 5.2 \\ 0.7\overline{)3.64} \\ \underline{35} \\ 14 \\ \underline{14} \\ 0 \end{array}$$

B. 5.2

6. $1\frac{2}{3} + 2\frac{5}{6} = 1\frac{4}{6} + 2\frac{5}{6} = 3\frac{9}{6} = 4\frac{3}{6} = 4\frac{1}{2}$

B. $4\frac{1}{2}$

7. $3\frac{2}{5} - 1\frac{1}{2} = 3\frac{4}{10} - 1\frac{5}{10} = 2\frac{14}{10} - 1\frac{5}{10} = 1\frac{9}{10}$

C. $1\frac{9}{10}$

8. $\frac{4}{5} \times 3\frac{1}{3} = \frac{4}{\cancel{5}_1} \times \frac{\cancel{10}^2}{3} = \frac{8}{3} = 2\frac{2}{3}$

C. $2\frac{2}{3}$

9. $2\frac{2}{5} \div 1\frac{1}{2} = \frac{12}{5} \div \frac{3}{2} = \frac{\cancel{12}^4}{5} \times \frac{2}{\cancel{3}_1} = \frac{8}{5} = 1\frac{3}{5}$

A. $1\frac{3}{5}$

10. $\frac{60}{84} = \frac{\cancel{2} \cdot \cancel{2} \cdot \cancel{3} \cdot 5}{\cancel{2} \cdot \cancel{2} \cdot \cancel{3} \cdot 7} = \frac{5}{7}$

C. $\frac{5}{7}$

11. B. 4

12. A. 2,500,000

13. $$1.2 \text{ m} \cdot \left(\frac{1 \text{ km}}{1000 \text{ m}}\right) = 0.0012 \text{ km}$$
$$1.2 \text{ m} \cdot \left(\frac{100 \text{ cm}}{1 \text{ m}}\right) = 120 \text{ cm}$$
$$1.2 \text{ m} \cdot \left(\frac{1000 \text{ mm}}{1 \text{ m}}\right) = 1200 \text{ mm}$$

C. 1200 mm

14. B. 2 m

15. $$\begin{aligned} 0.4 \times 100\% &= 40\% \\ 2 \div 5 \times 100\% &= 40\% \\ 40 \div 100 \times 100\% &= 40\% \\ 0.04 \times 100\% &= 4\% \end{aligned}$$

D. 0.04

16. $10^3 \cdot 10^4 = 10^{3+4} = 10^7$

B. 10^7

17. $\sqrt{100} = 10$, so $\sqrt{199} > 10$.
$14 \times 14 = 196$
$15 \times 15 = 225$

C. 14 and 15

18. Twenty-five thousandths = 0.025

E. None correct

19. $17.36 + 8.7 \approx 17 + 9 = 26$

D. 26

20. $33\% = 0.33$
$33\frac{1}{3}\% = \frac{1}{3}$
$\frac{1}{3} > 0.33$

$$\begin{array}{r} 0.333 \\ 3\overline{)1.000} \\ \underline{9} \\ 10 \\ \underline{9} \\ 10 \\ \underline{9} \\ 1 \end{array}$$

C. 0.3, 33%, $\frac{1}{3}$

21. $$\begin{aligned} \frac{8}{12} &= \frac{12}{n} \\ 8n &= (12)(12) \\ 8n &= 144 \\ n &= \frac{144}{8} \\ n &= 18 \end{aligned}$$

C. 18

22.

	Ratio	Actual Count
Boys	2	*B*
Girls	3	18
Total	5	*T*

$$\frac{3}{5} = \frac{18}{T}$$
$$3T = 90$$
$$T = 30$$

D. 30

23. $3(25) = 75$
$75 - 22 - 22 = 75 - 44 = 31$

D. 31

24. The measures, in order, are 6, 8, 8, 8, 9, 9, 10, 11, 12, 16. Because the number of measures is even, the median is the average of the two middle values:

$$\frac{9 \text{ in.} + 9 \text{ in.}}{2} = \frac{18 \text{ in.}}{2} = 9 \text{ in.}$$

B. 9 in.

25. 3 favorable outcomes
6 total outcomes

So, the probability is $\frac{3}{6}$, or $\frac{1}{2}$.

C. $\frac{1}{2}$

26.
$$\begin{array}{r} \$2.89 \\ \times\ 0.08 \\ \hline 0.2312 \end{array} \longrightarrow \$0.23 \text{ tax} \qquad \begin{array}{r} \$2.89 \\ +\ \$0.23 \\ \hline \$3.12 \end{array}$$

B. $3.12

27.
$$\begin{array}{r} 0.13 \\ 32\overline{)4.16} \\ \underline{3\,2} \\ 96 \\ \underline{96} \\ 0 \end{array}$$

C. 13¢/oz

28.

	%	Actual Count
Original	100	$36
Change	25	*C*
New	75	*S*

$$\frac{100}{75} = \frac{\$36}{S}$$
$$100S = \$2700$$
$$S = \$27.00$$

C. $27.00

29.
$$\begin{array}{r} 0.84 = 84\% \\ 25\overline{)21.00} \\ \underline{20\,0} \\ 1\,00 \\ \underline{1\,00} \\ 0 \end{array}$$

B. 84%

30. $400 \times \left(2\frac{1}{2}\right) = 1000$

D. 1000 miles

31. **B.**

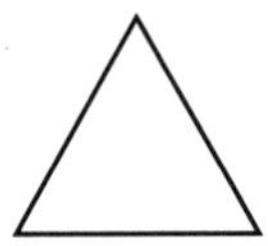

32. $m\angle x + m\angle y + 90° = 180°$
$m\angle y = 180° - 90° - m\angle x$
$m\angle y = 90° - 38°$
$m\angle y = 52°$

A. 52°

33. Perimeter = 7 cm + 11 cm + 8 cm + 9 cm + 15 cm + 20 cm = 70 cm

D. 70 cm

34. Circumference = $2\pi(r) \approx 2(3.14)(10 \text{ inches})$ = 62.8 in.

B. 62.8 in.

35. Area $= \frac{1}{2}bh = \frac{1}{2}(9\text{ m})(12\text{ m}) = 54\text{ m}^2$

B. 54 m^2

36. Area $= \pi r^2 = \pi(4)^2 = 16\pi \approx 16(3.14) \approx 50\text{ ft}^2$

A. 50 ft^2

37. Volume $= lwh$
$= 8\text{ cm} \times 5\text{ cm} \times 2\text{ cm}$
$= 80\text{ cm}^3$

A. 80 cm^3

38. **C.**

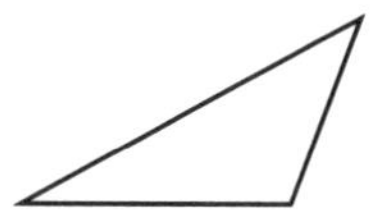

39. **D. $\overline{TS}$**

40. $3^2 + 4^2 = c^2$
$25 = c^2$
$5 = c$

A. 5

41. $(-6) + (-3) + (+2) = (-9) + (2) = -7$

B. –7

42. $4(3)(12) = 12(12) = 144$

D. 144

43. $3[16 - 2(5 - 3)] = 3[16 - 2(2)] = 3(16 - 4)$
$= 3(12) = 36$

C. 36

44.

```
      24
     /  \
    2    12
        /  \
       2    6
           / \
          2   3 ——> 2^3 · 3 = 24
```

A. $2^3 \cdot 3$

45. $6(25) = 6(20 + 5) = 6 \cdot 20 + 6 \cdot 5$

D. $6 \cdot 20 + 6 \cdot 5$

46. $3m - 1 = 35$
$3m = 36$
$m = \frac{36}{3}$
$m = 12$

B. 12

47. **A. $\sqrt{9} < 2^2$**

48. $(-10)(-10) = 100$

B. 100

49. Since $-3 < -2$, x cannot be -3.

D. –3

50.

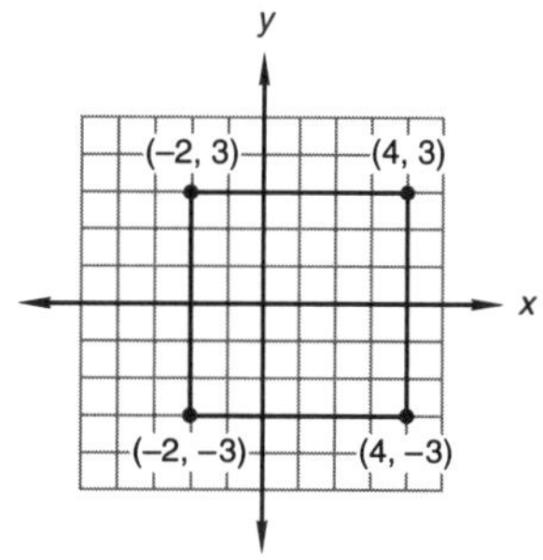

A. (–2, –3)